T0258243

The Foraging Strategy of Howler Monkeys
A Study in Primate Economics

THE FORAGING STRATEGY OF
HOWLER MONKEYS

A STUDY IN PRIMATE ECONOMICS

Katharine Milton

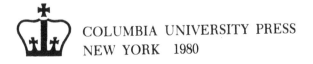

COLUMBIA UNIVERSITY PRESS
NEW YORK 1980

Library of Congress Cataloging in Publication Data

Milton, Katharine.
  The foraging strategy of howler monkeys.

  Bibliography: p.
  Includes index.
  1. Mantled howler monkey—Food. 2. Mantled howler
monkey—Behavior. 3. Mammals—Food. 4. Mammals—
Behavior. I. Title.
QL737.P925M54   599.8'2   79-27380
ISBN 978-0-231-04850-7

Columbia University Press
New York   Guildford, Surrey

For James Thomas Milton

CONTENTS

TABLES

FIGURES

ACKNOWLEDGMENTS

MOST MATERIAL presented here was collected to fulfill the requirements for my doctoral dissertation. I thank Glenn Conroy, my
dissertation chairman, for his interest in my work and for all of
his help with this project. My appreciation too to Alison Richard,
who first suggested that I visit Barro Colorado, and to Ira Rubinoff
for encouraging me to study howlers there.

I owe a tremendous debt to the scientists associated with the
Smithsonian Tropical Research Institute as well as the many researchers who visited Barro Colorado during my study. Their stimu-
lating comments, questions, and suggestions were invaluable to this
research.

I particularly wish to thank Egbert Leigh and Stanley Rand for
their unflagging interest in the activities of howler monkeys as
well as their many helpful questions and suggestions with respect
to this research.

Robin Foster generously gave up many hours to aid me with plant
identifications as did Nancy Garwood and Carol Augspurger.

Marjorie Shepatin prepared the figures for this book while Arlee
Montalvo and members of the Graphics Section, Pan-Canal Company, helped ready the material for publication. I very much ap-
preciate this much needed assistance.

Joe Ito, of the Oregon Regional Primate Center, Beaverton, Oregon, drew the cover picture while Roy Fontaine, Michael May,
and George Angehr contributed photographs. Howler monkeys
are very difficult both to draw and to photograph and I am particularly grateful for these excellent illustrations.

For their help and encouragement I would also like to thank
Pedro Galindo, Ken Glander, David Janos, Teddy Kowal, Michael
May, James Mellett, Russell Mittermeier, Gene Montgomery, Martin
Moynihan, James Robertson, Nicholas Smythe, Frederick Szalay,
Arturo Tarak, John Terborgh, and Peter Van Soest.

The latter part of this study was supported by a grant from the Noble Fund to the Smithsonian Tropical Research Institute. Computer time on the Hewlett-Packard 9830 was provided by the Smithsonian Tropical Research Institute; Gene Montgomery and Don Windsor helped with the programming. Computer time on the CDC 6600 was provided by the Department of Anthropology, New York University; Bert Holland helped with the programming.

This study could not have been completed without the continuous interest and encouragement of Tom Milton. His assistance in every stage of this project is gratefully acknowledged and deeply appreciated.

The Foraging Strategy of Howler Monkeys
A Study in Primate Economics

1 PRIMATES AND PLANT FOODS: PROBLEMS AND SOLUTIONS

PRIMATES AND tropical forest trees share a long history of association. It is generally accepted that primates evolved in tropical forests where most primates are found today. Further, the great majority of primates are arboreal animals; therefore, it is not surprising that many primates are strongly dependent on plant foods, particularly seasonal dietary items such as the new leaves, flowers, and fruits of forest trees and vines. All extant primates, even the most minute prosimians, take at least some portion of their diet from plants, and most larger primates are entirely or almost entirely primary consumers.

At one time, it was popularly believed that tropical forests were stable and unchanging environments that offered almost unlimited food resources to plant-eating animals. Recent research, however, indicates that this is a misleading assumption. Tropical forests generally show considerable seasonal variability with respect to phenological patterns. At certain times of year, many tree species produce an abundance of new leaves, flowers, or fruit; at other times, however, there are notable and persistent lows in the production of such items and concomitant lows in their availability as foods for primates.

Further, there has been a growing awareness that trees and other plants do not passively accept the destruction and removal of their leaves, flowers, and unripe fruits by herbivores but rather have a variety of strategies for counteracting or minimizing such damages. For example, the staggered and varied phenological patterns of many tropical forest tree species appear to function, at least in part, to minimize or discourage herbivore pests. Some tree species produce new leaf crops at times of year when particular herbivores are at low densities; others produce new leaf crops at a time of year when many other tree species are also producing them. Thus potential herbivores are faced with a sudden superabundance of resources that rapidly disappear as the leaves mature. In addition, trees protect their vegetative and reproductive parts with a wide variety of toxic or distasteful chemicals as well as mechanical devices such as thorns or hairs. Most tree species use various of these strategies simultaneously, in effect erecting a series of barriers against herbivore attack.

To exploit plant foods in a tropical forest, animals must therefore have some method of tracking these illusive and ephemeral resources through space and time and a way of dealing with their variable chemical constituents, both nutritional and toxic. This may require dietary strategies of considerable complexity. Primates, like other animals, are seeking to optimize their net returns from foraging. If too much time and energy are expended in searching for these variable and patchily distributed plant foods, the costs of foraging may well exceed the benefits. Thus primates dependent on plant foods, particularly larger primates that require more food and larger supplying areas than small primates, must develop efficient strategies for locating preferred foods. In the pages that follow, I describe the foraging strategy of one highly successful arboreal primary consumer, the mantled howler monkey (*Alouatta palliata*), and show how it has been able to overcome many of the problems inherent in a diet composed entirely of the leaves, flowers, and fruits of tropical forest trees and vines.

BACKGROUND

Howler monkeys (Family, Cebidae; Subfamily, Alouattinae Elliot 1904; Genus, *Alouatta* Lacépède 1799) are the largest of the New World primates, averaging some 7 kg to 9 kg when adult. They are highly arboreal, generally confining their activities to the middle and upper branches of tall forest trees and only rarely coming to the ground (Carpenter 1934). Howlers are found in a diverse array of forest types, including lowland evergreen forest, highland forest, swamp forest, and riparian forest. Their overall geographical range extends from southern Mexico, through much of Central and South America, and into northern Argentina. In sites where this has been measured, howlers appear to form the highest percentage of the overall primate biomass—an important criterion of ecological dominance (Eisenberg and Thorington 1973). Six *Alouatta* species are currently recognized: *A. belzebul* Linnaeus 1766, *A. seniculus* Linnaeus 1766, *A. caraya* Humboldte 1812, *A. palliata* Gray 1849; *A. fusca* Ihering 1914 and *A. pigra* Lawrence 1933 (Hill 1962; Hershkovitz 1969; Smith 1970).

Most descriptions of howler monkeys have stressed the fact that their diet typically includes a substantial proportion of leaves (Carpenter 1934; Fooden 1964; Hladik and Hladik 1969; Richard 1970; Smith 1977). Yet leaves are a notoriously poor source of ready energy and generally contain high proportions of relatively indigestible structural materials as well as potentially harmful chemical compounds such as alkaloids and phenolics. Animals that specialize on leaves in the diet, even for only short periods of time each year, must therefore have behavioral and/or morphological features to help overcome some of the problems inherent in using leaves as food. Some Old World leaf-eating primates, for example (i.e., Colobinae and Indriidae), have specialized sections in the digestive tract that enable them to degrade plant structural carbohydrates through bacterial fermentation (Bauchop and Martucci 1969; Bauchop 1971). Studies of the feeding behavior of certain of these Old World folivores have shown that at times substantial proportions of young leaves, shoots, and fruits may be included in the diet but at other times such animals may feed entirely or almost entirely on mature leaves and even bark and dead wood (Yoshiba 1968; Hladik and Hladik 1972; Clutton-Brock 1974; Richard 1977). They are presumably able to exploit such fibrous foods efficiently because of the nutritive benefits provided by cellulolytic gut flora.

At the time I began this study (March 1974), the prevalent view in the literature was that howler monkeys were New World analogues to these highly specialized Old World forms (Hill 1962;

Hladik and Hladik 1969; Richard 1970; Eisenberg et al. 1972; Jolly 1972; Eisenberg and Thorington 1973). This implied that howlers had digestive specializations of the same type and/or magnitude as colobines and indriids and occupied a similar dietary niche in forests of the New World. Yet there had been little attempt to compare the digestive tract of howlers with those of Old World folivores, nor had there been any long-term quantitative study of howler feeding ecology. Almost nothing was known about the relative nutritional components of tropical tree foliage from either the New or Old World nor about its presumed chemical defenses. A number of researchers had studied specific aspects of howler behavior, but quantitative data on feeding behavior and diet were scarce and occasionally contradictory. In particular, except for the work carried out by the Hladiks and their associates (1969, 1971), there had been no attempt to examine howler feeding behavior in relation to the density, distribution, and phenological patterns of potential food sources.

For all of these reasons it seemed desirable to instigate a long-term study of howler foraging behavior in a tropical forest habitat, particularly since it appeared that the set of adaptations to leaf-eating by howler monkeys might well differ in certain important respects from those of specialized Old World primate folivores.

PROBLEMS OF PRIMARY CONSUMERS

Food is obviously of critical importance to all animals and selective pressures related to diet should have a profound influence on many aspects of an animal's behavior, morphology, and physiology. In recent years, there has been a growing interest in the strategies animals use to meet their nutritional requirements. This has resulted in a rapidly growing body of information that is known collectively as foraging strategy theory (MacArthur and Pianka 1966; Emlen 1966; Schoener 1971). A foraging strategy may be defined as those aspects of behavior and morphology of an animal that are involved in the procurement and utilization of food (Schoener 1971). Foraging theory attempts to explain and ultimately to predict how a given animal will respond to the potential food resources within a given habitat. The theory postulates that the efficiency of foraging—as measured in net energy yield/foraging time or some other units that may reflect fitness—is maximized by natural selection (Schoener 1971). Many mathematical models used in the theory seem to have been formulated mainly with secondary consumers in mind, and some researchers have pointed out the difficulty in applying such models to primary consumers.

In 1974, for example, Westoby proposed that, in view of the relatively low nutritional value of most plant foods and the limits of an animal's capacity to process food, the objective of a large generalist herbivore should be to optimize the nutrient mix within a given total bulk of food, rather than maximize the energy yield/foraging time. He concluded that since the foraging activities of herbivores are limited by digestion time, rather than by search or pursuit time, food *quality* should be more important than availability in determining food selection. Freeland and Janzen (1974), who also considered the problems of food selection by generalist herbivores, suggested that choice of diet may be affected by the presence of potentially harmful secondary compounds in most plant parts. They hypothesize, as Westoby did, that herbivores should be selective in their feeding behavior with regard to specific classes of nutrients and to secondary compounds.

Being selective in feeding, however, would increase the cost of food procurement, since it would increase the amount of time spent searching for food (i.e., traveling). It would also increase the risk of exposure to predators. If the efficiency of foraging is maximized by natural selection, as postulated by the theory of foraging strategy, then primary consumers should have behavioral and morphological features that function to minimize the costs of procuring preferred foods. Such features should be most evident in the activity of food location (search strategy), since the foods eaten by primary consumers are sessile and relatively predictable, as compared with the foods eaten by secondary consumers. Which type of search strategy is most efficient depends on the distribution of the preferred foods. It will therefore be useful, before going further, to consider some aspects of the content and distribution of the foods eaten by primary consumers in a tropical forest.

Food Content

Some models of foraging strategy assume that different foods have equal values in terms of energy and nutrients (e.g., MacArthur and Pianka 1966); others assume that the value of the food can be measured in terms of a single nutrient, usually energy (e.g., Emlen 1966). Most of them also assume that any food in hand is worth eating. It may be that certain classes of animal prey are not highly variable in energetic and nutritive value, and that they contain similar proportions of energy and nutrients (Maynard and Loosli 1969). It may also be that any such prey in hand is worth eating. But these assumptions certainly do not hold for the plants and plant parts that make up the diets of primary consumers.

BULK. Plant parts generally contain from 60 percent to more than 90 percent water (Hladik et al. 1971; Casimir 1975; Milton, unpub.). Further, many plant parts, particularly leaves, have high concentrations of structural materials—from 30 percent to 60 percent or more dry weight (Milton 1979a)—that cannot be efficiently degraded by animals lacking an extensive gut flora (Moir 1965; Van Soest 1980). Thus food from plants offers a relatively low return of nutrients for its bulk. Primary consumers, which face constraints with respect to the amount of food they can process per unit time (Bell 1971; Westoby 1974), must select items that offer the best net return for the time and energy spent in finding, eating, and digesting them. Many plant foods may actually not be worth eating—i.e., the net return is zero or even negative (see, for example, Sinclair 1974, 1977; Milton 1979a).

VARIABILITY. Leaves from different plant species and even leaves from the same species at different stages of maturity or at different times of day may vary considerably in their nutritional and/or toxic content (Feeny 1970; Rockwood 1974; Casimir 1975; Milton 1979a). For example, mature leaves from certain tree species on Barro Colorado Island, Panama Canal Zone, were found to contain an average of 9.3 percent protein per unit dry weight while immature leaves of the same species averaged 12.4 percent (summed individual amino acids; Milton 1979a). In cases where mature and immature leaves contain similar amounts of protein, mature leaves might still be less desirable as a food source since they generally contain higher proportions of structural materials than immature leaves (*ibid.*). Feeny (1970), for example, found that mature oak leaves were over seven times as tough as immature oak leaves and had lower protein concentrations. He also found that mature oak leaves had higher concentrations of proteinase-inhibiting tannins than immature oak leaves, and McKey (1974, 1978) and Rhoades and Cates (1976) have suggested that mature leaves of trees in general may have higher concentrations of certain secondary compounds than immature leaves. Fruits and flowers also vary widely in both palatability and nutritional content (Hladik et al. 1971; Hladik 1977a, 1977b, 1978).

INCOMPLETE NUTRIENTS. Fruits may be rich in nonstructural carbohydrates or lipids but are generally poor in protein; leaves, particularly young leaves, may contain considerable protein, but are generally very low in nonstructural carbohydrates and lipids (Hladik and Hladik 1969; Hladik et al. 1971; Hladik 1977a, 1977b; Casimir 1975; Milton 1979a). It may therefore be necessary for many primary consumers to eat foods from both of these categories each day to get the required balance of ready energy and protein. Further, the protein in some plant foods is incomplete, i.e., one or more essential amino acids is lacking or present only in trace amounts (Guthrie 1971). It may therefore be necessary for many primary consumers to eat foods from various sources in order to get all of the essential amino acids in the required amounts and proportions. In ruminants, however, and in other animals with similar digestive specializations, certain essential amino acids can be synthesized by gut flora (Allison 1965; Moir 1965, 1967; Hungate 1967; Bauchop and Martucci 1969).

SECONDARY COMPOUNDS. All plant parts contain secondary compounds, some of which may function to protect them from herbivore predation (Fraenkel 1959; Ehrlich and Raven 1965; Whittaker 1970; Feeny 1970; Levin 1971, 1976). Certain secondary compounds are distasteful or malodorous; others may be toxic or even fatal to the feeder. Some secondary compounds, particularly condensed tannins, can bond with protein in the gut and inhibit the action of digestive enzymes (Feeny 1970; Freeland and Janzen 1974; Ryan and Green 1974). Toxic substances ingested with food must be neutralized and removed from the body, i.e., they must be detoxicated, a process which can be energetically expensive as it may require products of primary metabolism to conjugate with the toxins so that they can be excreted (Parke 1968; Williams 1969). One might expect to find higher concentrations of toxic or proteinase-inhibiting compounds in leaves than in ripe, pulpy fruit, as fruit pulp often appears to function as an attractant for seed-dispersal agents (Morton 1973; McKey 1975). Thus, the more an animal depends on leaves as a food source, the more of a problem it may have with secondary compounds.

FOOD SELECTION. In view of the above factors, primary consumers should be selective in their feeding behavior with respect to nutritional quality, nutrient mix, and secondary compounds, as predicted by Westoby (1974) and Freeland and Janzen (1974).

Food Distribution

Primary consumers face problems not only with the variable nutritional content of potential foods in a tropical forest, but also with the distribution patterns of such foods. At first glance a tropical forest may appear to offer omnipresent and abundant sources of food, particularly to leaf-eating animals. A closer examination, however, reveals many fallacies in this assumption.

SPECIES DIVERSITY. Tropical forests typically have a high diversity of tree species, none of which is dominant (Richards 1952). Thus the probability of encountering a given species in such a forest is relatively low. This has obvious implications for the problem of food location. It suggests that for an animal as large as an adult howler monkey (7 kg to 9 kg), the cost of restricting its diet to one or even a few species would be prohibitively high since such an extensive supplying area would be required.

SEASONAL VARIABILITY. The concept of tropical forests as stable and unchanging environments has been shown to be a considerable oversimplification. Many tropical forest areas have clear fluctuations in the amount of rainfall received at different times of the year, typically referred to as wet and dry seasons (Smythe 1970; Frankie et al. 1974; Leigh 1975; Leigh and Smythe 1978). This affects the phenology of much of the vegetation (Richards 1952; Janzen 1967; Fogden 1972; Frankie et al. 1974; Jackson 1978). Studies have shown that many tropical tree species have particular phenological patterns that appear related to climatic factors (Janzen 1967; Smythe 1970; Foster 1973; Augspurger 1978). Many tree species on Barro Colorado Island, for example, maximize fruit production in the late dry and mid-rainy seasons (Smythe 1970; Foster 1973). This is not to imply that all tree species follow the same phenological patterns with respect to leaf, flower, or fruit production, but a great many do, with the result that there are clear peaks and valleys in the overall production of these seasonal items by forest trees. Thus a primary consumer in a tropical forest may be faced with periods when certain foods are relatively scarce (Fogden 1972; Foster 1973; Leigh 1975).

SYNCHRONY IN PHENOLOGY.   Most Neotropical tree species show at least some degree of intraspecific synchrony in phenology (Richards 1952; Croat 1967, 1978; Frankie et al. 1974; Augspurger 1978), with individuals of species in a given area going through the same phenological phase (i.e., flowering, fruiting, flushing) only during a certain time span. The effect of this on a primary consumer is that a particular food item may be available in large amounts during a certain period but it is not available at all during the rest of the year. Thus, in most cases, an animal cannot depend on trees of the same species to provide a continuous supply of new leaves, flowers, or fruit. Nor can animals depend on a particular individual of a given species to produce such crops each year. Individual trees, for example, may produce a massive fruit crop in one year and then skip a year or more before producing another. Most members of a species may also follow this pattern (Milton unpub.).

EPHEMERALITY OF POTENTIAL FOOD ITEMS.   Young leaves, fruits, and flowers are typically available on individual trees for very short periods of time. In the case of leaves, it is presumably not advantageous for most plants to delay maturation, since until leaves reach a certain stage of maturity they are net users rather than providers of energy. Nor is it advantageous for a plant to lose most of its new leaves to herbivore predators. It has been suggested that one defensive strategy of certain tree species is to produce new leaves synchronously when herbivore populations are at low levels or inactive and to mature them before herbivores can build up again (Varley and Gradwell 1962; Rockwood 1974; McKey 1975). In this strategy, survival of the leaf would depend on rapid maturation. The leaf is thereafter apparently protected by its greater toughness and/or heavier concentrations of toxic compounds (Feeny 1970; Rockwood 1974). Further, there is evidence which suggests that in some species, once an herbivore predator has eaten a portion of a new leaf crop, the plant responds within twenty-four hours by increasing the defensive compounds (proteinase inhibitors) in its leaves (Ryan and Green 1974; Hankioja and Niemala 1975). Thus, at least in some species, the young leaves on an individual plant may be optimally edible for as little as one day.

Fruit may also be optimally edible for very short periods of time. Green fruit is typically protected from predators by a variety of chemical compounds, many of which are highly astringent and distasteful; and after the point when maturation is complete, the climacteric, fruit begins to deteriorate—i.e., the sugars, among other substances, begin to decline (Spencer 1974). Ripe fruit, on individual plants, may thus be optimally edible for no more than a few days.

Flowers are generally even more ephemeral. As an extreme example, flowers appear on some species of Neotropical Myrtaceae for only one day a year and are withered by the following day (Richards 1952). At the other extreme, some species are characterized by individuals that produce at least a few flowers almost continuously over a period of months (*Pseudobombax septenatum, Tabernaemontana arborea*). Whatever the strategy, in most cases an individual flower appears optimally edible for no more than a day or two.

TWO BASIC STRATEGIES

Given these problems, one possible strategy for a primary consumer in a tropical forest would be to use a food resource that provides most of the essential nutrients and is continuously available in space and time. Mature leaves are such a resource, since they would provide (directly and indirectly) both protein and ready energy if the animal were able to degrade cellulose and hemicelluloses through bacterial fermentation. But to utilize mature leaves efficiently, the animal may require elaborate digestive specializations (e.g., an enlarged sacculated stomach or an extensive caecum and/or colon). This does not mean that such an animal would depend exclusively on mature leaves; for all of the reasons given above, it would still be selective in its feeding behavior. But it could afford to be *less selective* than an animal lacking such specializations and, when higher quality foods were scarce, it could subsist on a diet with a large proportion of mature leaves.

Another possible strategy would be to become a *very* selective feeder and to depend mainly on higher quality foods (e.g., young leaves, fruits, and flowers)—which may always be available from some species in a tropical forest—and to develop behavioral and morphological features that would minimize the costs of procuring such foods.

A comparison of certain features of the digestive morphology, as well as the available quantitative data on diet, suggests that Old World colobines and indriids may have evolved the first basic strategy and that howlers may have evolved the second.

Colobinae and Indriidae

COLOBINAE.   Colobine monkeys differ from all other primates in the large size and anatomical complexity of their stomachs (Bauchop and Martucci 1968; Bauchop 1971, 1978). As pointed out by Bauchop (1971), these differences are related to a diet consisting at times primarily of leaves, hence the term "leaf-eater" commonly used to describe these primates. The colobines have sacculated stomachs similar in many respects to the chambered stomachs of ruminants. The structure of the stomach permits separation of the

ingesta in the proximal parts from the distal acid pyloric region.
The large capacity of the stomach allows the accumulation of ingesta and the slow rate of passage essential for the efficient fermentation of plant materials.

Gastric samples obtained before the morning feeding of *Presbytis cristatus* and *P. entellus* (10.5 hours after the previous feeding)
showed a continued high production of volatile fatty acids (VFA's).
These fatty acids presumably make an important contribution to
the energy metabolism of such primates and may also benefit their
vitamin and nitrogen economies (Bauchop and Martucci 1968; Bauchop 1971; Parra 1978). Further, it has been shown that some types
of gut flora perform detoxicatory functions, which would reduce
the demands on the microsomal enzymes of host animals (Schuster
1964; Scheline 1968; Williams 1969, 1971).

INDRIIDAE. The indriids have simple but capacious stomachs
(Flower 1872; Hill 1953). However, they have greatly elongated
caeca and colons which occupy the entire posterior region of the
abdomen. The lower intestine is remarkable for its great length; in
Propithecus it is nine times total body length, in *Avahi* fourteen
times total body length, and in *Indri* fifteen times total body length
(Hill 1953). Further, the length of the caecum in all mature indriids
exceeds total body length and the caecum is greatly sacculated as
well (Hill 1953; Hladik 1967).

There is no body of work on the digestive processes of indriids
similar to those cited above for colobines. It would seem that the
capacious stomach of indriids enables them to take in large quantities of bulky, high fiber foods. Most soluble nutrients are presumably extracted in the stomach and small intestine while the fibrous
material is passed into the highly specialized lower intestinal tract.
Here there should be sufficient volume to retain quantities of this
material for sufficient time to permit the extensive bacterial activities necessary to utilize such foods efficiently.

DIET. Few Old World leaf-eating primates have been studied
in sufficient detail so that the actual proportions of different items
of diet are known. In all species that have been so studied, however, there are periods in an annual cycle when perennial foods
(e.g., mature leaves and/or bark) are a major dietary component.

In some weeks, *Propithecus verreauxi* has a diet consisting of as much as 70 percent mature leaves and/or 20 percent dead wood (Richard 1973); Yoshiba observed that in the dry season in Dharwar, bark was an important dietary component for *Presbytis entellus* (1968); both the red colobus (*Colobus badius*) and the purple-faced langur (*P. senex*) have diets consisting at times of over 60 percent mature leaves (Hladik and Hladik 1972; Clutton-Brock 1974, 1975); in some weeks, *C. guereza* lives almost exclusively on the mature foliage of only one tree species (Clutton-Brock 1974).

Members of both the colobines and indriids apparently prefer to feed on more succulent younger foliage as well as fruits and flowers when they are available. When such foods are in short supply, however, species of both groups apparently can turn to mature leaves, bark, and other fibrous perennial foods to carry them through.

Howler Monkeys

DIGESTIVE TRACT. Fermentation of plant fiber generally takes place either in the stomach or in the caecum and/or colon. Howler monkeys have a simple, unsacculated stomach (Hill 1962; Cramer 1968; Milton, unpub.). It is capacious but, as the comparative charts of Fooden (1964) and Hladik (1967) clearly show, it has no more relative surface area—and often has less—than the stomachs of many other primates not generally regarded as leaf eaters (table 1.1). For example, the stomach of *Chiropotes*, a highly frugivorous monkey (Fooden 1964), has more relative surface area than that of *Alouatta;* however, the stomach of *Colobus polykomos*, a specialized Old World leaf eater, has a relative surface area some six times that of *Alouatta.*

The caecum of the howler has somewhat more relative surface area than those of many primates not generally regarded as leaf eaters; however, the difference in area is often slight. *Ateles paniscus*, a highly frugivorous primate, has only slightly less caecal area than *Alouatta*. The caeca of more specialized Old World leaf eaters, however, have considerably more relative surface area than that of *Alouatta*. The colon of howlers, another potential fermentation site, is notably smaller than those of many fruit-eating primates such as chimpanzees and mangabeys, as well as specialized

TABLE 1.1. A Comparison of the Relative Surface Area of Different
Sections of the Digestive Tracts of 24 Primate Species

	N		Relative Surface Area		
		Stomach	Small Intestine	Caecum	Colon
Arctocebus calabarensis	1	30	300	20	60
Perodicticus potto	1	30	200	15	140
Galago demidovii	4	44	275	35	85
Galago elegantulus	1	65	160	45	140
Galago alleni	3	58	317	27	135
Microcebus murinus	1	45	300	15	55
Cheirogaleus major	4	56	290	14	66
Leontocebus midas	1	20	195	10	40
Cebus griseus	2	35	308	3	25
Ateles paniscus	1	60	280	35	80
Papio leucophaeus	2	15	145	15	155
Papio papio	1	25	200	5	80
Cercocebus albigena	3	32	190	22	217
Cercopithecus nictitans	4	40	224	24	189
Cercopithecus neglectus	1	70	260	35	340
Cercopithecus vervet	1	25	120	10	100
Cercopithecus pygerythrus	1	10	150	20	65
Cercopithecus talapoin	4	55	225	25	165
Gorilla gorilla	1	40	220	15	190
Pan troglodytes	1	70	250	25	260
Mean		41	230	21	129
Range		10–70	120–317	3–45	25–340
Alouatta seniculus	1	50	300	40	160
Colobus polykomos	1	320	180	10	180
Avahi laniger	3	92	547	253	280
Lepilemur mustelinus	1	70	105	270	250

Source: All data except for means and ranges come from the work of C. M. Hladik
(1967).
Note: Relative surface area = (actual surface area)/(length of body + head)^2 ×
1000.

Old World forms. Indeed, the total length of the intestinal tract of
Alouatta is described by Hill (1962) as "remarkably short" for even
a fruit- and leaf-eating monkey.

It thus appears that the digestive tract of howlers does not have
specialized features of the same type and/or magnitude as those of
colobines and indriids. However, this does not preclude the possibility of fermentation activities in howlers for various monogastrics

with relatively simple hindguts are known to ferment fibrous plant materials (Keys et al. 1969). Nor does this imply that gut flora might not aid in detoxication, though this possibility appears unlikely. But the magnitude of the differences between the howler digestive tract and those of Old World leaf eaters strongly suggests that the efficiency of such processes is considerably lower in howlers (see chapter 6).

There are few data available on the relative abilities of any primate species with respect to degrading plant structural carbohydrates (Bauchop and Martucci 1968; Bauchop 1971; Spiller and Amen 1975; see Bauchop 1978 and Parra 1978 for a review of information to date). It is highly probable that *all* primates, including man, are able to digest some cellulose and hemicelluloses (Spiller and Amen 1975; Stevens, personal communication; Van Soest, personal communication). The critical difference between primates must be in their *relative efficiencies* at such processes. Primates such as colobines and indriids should excel at degrading structural carbohydrates since they have the capacity to maintain huge bacterial colonies and hold bulky foods in the digestive tract for a sufficient amount of time to utilize them efficiently.

THE HOWLER DIET. The literature generally describes howlers as digestively specialized leaf-eating primates that also eat fruit (Carpenter 1934; Fooden 1964; Hladik and Hladik 1969; Richard 1970; Eisenberg et al. 1972; and others). Though there seems to be general agreement that howler monkeys are more folivorous than other cebids, the literature is ambiguous as to how folivorous they actually are. For example, various researchers on the howler diet on Barro Colorado Island (the principal research site for all earlier studies of howler behavior) agree that howlers eat leaves but estimates range from as little as 5 percent leaves and 95 percent fruit (Altmann 1959), to 50 percent leaves and 50 percent fruit (Richard 1970), to 40 percent leaves and 60 percent fruit (Hladik and Hladik 1969). Though some researchers mention that the howler diet includes mature leaves (Hladik and Hladik 1969), most accounts do not differentiate the age of leaves eaten. Since these studies were all carried out at the same study site (table 1.2), the differences in the estimates of the relative importance of leaves and fruits in the

TABLE 1.2. Field Research on Barro Colorado Howler Monkeys, 1932–1976

Researcher and Years of Field Work	J	F	M	A	M	J	J	A	S	O	N	D	Area Observed	Troops Observed and Type of Study
Carpenter 1931–1933 1959	+ +	+ +	+	+	+							+ +	Lutz Ravine	One troop irregularly observed; island-wide census
Collias and Southwick 1951	+	+	+		+	+	+						Lutz Ravine	One troop briefly observed; island-wide census
Altmann 1955	+	+	+	+							+		Lutz Ravine	One troop; general behavioral data.
Bernstein 1962	+ +	+ +											Lutz Ravine	Two troops; general behavioral data.
Chivers 1967						+	+	+					Lutz Ravine	Various troops; partial census, behavioral data, esp. vocalizations.
Hladik and Hladik Nov. 1966– Jan. 1968	+	+	+	+	+	+	+	+	+	+	+	+	Lutz Ravine	Various troops; dietary data.
Smith 1967–68	+	+	+	+	+	+	+	+	+	+	+	+	Lutz Ravine	Various troops; general observations, particularly dietary data.
Richard 1970						+	+	+					Lutz Ravine	One troop; activity patterns of howlers and spider monkeys.
Mittermeier 1973									+		+		Lutz Ravine	Six troops; general behavioral data.
Milton 1974–1976, on-going...	+	+	+	+	+	+	+	+	+	+	+	+	Lutz Ravine and Old Forest	Two troops; dietary data, foraging behavior and general activities, particularly ranging patterns.

howler diet doubtless reflect the fact that different researchers worked on the island at different times of the year, used different sampling techniques, and were there for different amounts of time. This lack of agreement strongly suggested to me that howlers might well vary the percentage of time they spent eating leaves, depending on the season of the year or some other factors.

The leaves and fruits eaten by howlers on Barro Colorado have been described as abundant, readily available, and easily obtained (Carpenter 1934; Hladik and Hladik 1969; Smith 1977; and others). Hladik and Hladik (1969) suggested that the bulk of the howler diet came from a few common tree species. Howlers have been said to remain in a single area of their home range for several days, moving little, and heavily exploiting all available resources before moving on to a new area (Carpenter 1934). It should be noted that these descriptions of the abundance and ease of obtaining potential foods in the Barro Colorado forest are at variance with the predictions of Westoby (1974) and Freeland and Janzen (1974) regarding the probable foraging behavior of generalist primary consumers; they are also at variance with material presented above on the difficulty of obtaining an adequate dietary mix due to the variable nutritional content and patchy distribution of potential foods in a typical tropical forest.

The nutritional content of the howler diet on Barro Colorado has been described as "poor" and high in cellulose for a monogastric (Hladik and Hladik 1969). Unfortunately few supporting data are available to demonstrate that the diet of howlers on Barro Colorado is any poorer in quality than diets of many other primates. Further, Smith (1977) has stated that howlers are very poor assimilators of their diet. Again, there are few supporting data to show that howler monkeys are any poorer at nutrient assimilation than many other primate species.

From these various descriptions, it is difficult to form a clear view of the howler diet on Barro Colorado, but one may conclude that howlers (a) have digestive specializations for leaf-eating, (b) are rather specialized in diet and take the bulk of their food from a few common tree species, (c) eat foods that are abundant and readily available, including considerable proportions of mature leaves, (d) have a diet low in nutritional value, and (e) are poor assimilators of this diet. Such a view is actually fairly coherent except, perhaps, for (e), but it depends heavily on the implicit assumption that howlers have gastro-intestinal specializations of a similar type and/or magnitude as those of specialized Old World primate folivores. If this were the case, howlers might well be able to be less selective in their choice of dietary items since they could depend heavily on the products of bacterial fermentation to augment nutritional deficiencies in their foods. Since the available morphological data indicate that this assumption is probably invalid, the whole above view of the howler diet and feeding behavior is open to question, especially as it is largely based on data collected during short-term studies in one small area of Barro Colorado Island (see table 1.2).

RESEARCH EXPECTATIONS

Given the importance of food content to primary consumers, as pointed out by Westoby (1974), Freeland and Janzen (1974), and others, and the apparent lack of digestive specializations of the same type and/or magnitude as those of Old World colobines and indriids, I expected howlers to be very selective in their feeding behavior with respect to nutritional quality, nutrient mix and secondary compounds. But, given the theory of foraging strategy (Schoener 1971) and what I knew about the patchy distribution of potential foods for primary consumers in a tropical forest, I expected howlers to have behavioral and morphological features that would minimize the costs of procuring preferred foods. Thus, I expected the foraging strategy of howlers to be a compromise between the pressure to be selective in food choice and the pressure to minimize the costs of food procurement.

2 THE GENERAL RESEARCH PLAN

To DETERMINE features characteristic of the foraging strategy of howler monkeys, I carried out a fourteen-month study of howler feeding ecology on Barro Colorado Island, Panama Canal Zone, where most previous field work on howler behavior had been done (Carpenter 1934; Collias and Southwick 1952; Altmann 1959; Bernstein 1964; Hladik and Hladik 1969; Chivers 1969; Richard 1970; Mittermeier 1973; Smith 1977).

Barro Colorado was created in 1914 by the damming of the Chagras River. This resulted in the formation of Lake Gatun, the principal water supply for the Panama Canal. Barro Colorado, the largest island in the lake, is approximately 15.5 km² in area and 4.8 km across at its widest diameter. The terrain is rugged, consisting of a series of low hills, the highest of which is 137 m above the level of the lake and 164 m above sea level.

The average annual rainfall on Barro Colorado is 2,730 mm (Croat 1978). Most rain falls from May to November, the "wet" season. This wet season is followed by a short "transition" season from mid-December to early January and then a pronounced "dry" season from mid-January through April. During the dry season, mean monthly rainfall is 39 mm. Data collected in the Barro Colorado area over the past thirty-eight years (Croat 1978) show that these seasonal rainfall patterns are highly predictable (figure 2.1), though the time of onset and the duration of the rainy season, as well as the amount of rainfall received, can show considerable annual variation.

Temperature, however, shows little annual variation and ranges between 23°C and 29°C the year round. Relative humidity varies

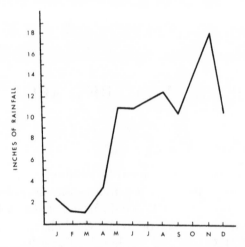

Figure 2.1.   Average Monthly Rainfall in Inches for the Barro Colorado Area
Source: Based on thirty-nine years of data collected by the Meteorological and Hydrographic Branch, Panama Canal Company and analyzed by T. Croat (1969).

between 50 to 100 percent in the dry season and 70 to 100 percent in the wet season (Allee 1926; Smythe 1970).

The forest on Barro Colorado was partially cut over in the early 1920s, but has been undisturbed now for over half a century. Virtually the entire island is covered with dense forest; many tree species are characteristic of mature forest, while most of the remainder are characteristic of late secondary growth conditions (Foster 1973; Knight 1975).

Howlers found on Barro Colorado are the descendants of animals occurring naturally in the area that were isolated when Lake Gatun was created. The density of howler monkeys on Barro Colorado (i.e., 1.13 animals per hectare; Milton 1977) is not a high density for this species. *Alouatta palliata* was reported to live at a density more than ten times this high in another area of Panama with no noted ill effects (Baldwin and Baldwin 1972). The Barro Colorado population was estimated to be around 450 animals in the early 1930s (Carpenter 1934). At this time, the howler population is estimated to be between 1,200 and 1,300 monkeys (Milton

1977) and evidence from several lines of investigation indicates that the population is no longer increasing in size (Smith 1977; Milton 1978c).

GENERAL DESCRIPTION OF STUDY SUBJECTS

Howler monkeys (*Alouatta palliata* Gray 1849) on Barro Colorado Island live in a number of discrete troops, each of which occupies a clearly defined home range. Home ranges of adjacent troops typically overlap. There is a strong and persistent tendency for howler troops on this island to average some 17 to 19 animals and to consist of 3 to 4 adult males, 5 to 7 adult females and 5 to 7 immature animals (Carpenter 1934, 1962; Milton 1975). Each troop is essentially a closed social unit and a high degree of interrelatedness appears to exist between its members. Typically howler troops show strong antipathy toward one another and toward any strange conspecific, but some movement of individuals between troops does occur. Adult males seem most likely to change troops (Carpenter 1934; Milton, unpub.).

As noted, howlers are highly arboreal, preferring the middle and upper levels of the forest canopy (Carpenter 1934; Mendel 1976). In some five years of work on Barro Colorado, I have never seen a howler monkey on the ground though they have been seen on the ground in other locales (Terborgh, personal communication). The locomotor mode of howlers is described as that of pronograde quadruped (Stern 1971). Howlers lope or walk quadrupedally along large horizontal branches and move at a cautious walk along narrow ones. When they reach the fine, terminal meshwork of branches between two tree crowns, they retain a grip with the hind limbs and tail, stretching the forelimbs forward to grasp the terminal branches of the adjacent tree. Once a firm grip is secured, the grip of the hind limbs and tail is released and the monkey passes into the next tree. Stern (1971) has pointed out that howlers typically move at a sedate pace. While this is generally the case, howlers are capable of very sudden, rapid movement and can run and leap as actively as any primate.

The most striking morphological adaptations of the howler appear to be its specialized laryngeal area and prehensile tail. The greatly enlarged hyoid bone enables howlers, particularly the male animals, to produce their loud, sonorous roar, which appears to function primarily as an inter-troop spacing mechanism (Chivers 1969; Marler 1970). The prehensile tail aids in locomotion and in maintaining balance when resting and sleeping. It is also used to help in harvesting many foods. Grand (1972) and Mittermeier and Fleagle (1976) have noted that suspensory postures can greatly increase the feeding sphere of an animal. Howlers typically sit or stand when feeding (Mendel 1976), but when preferred foods are out of reach, the animals suspend themselves by the tail and hind limbs or by the tail alone and harvest all edible items within the area covered by their hanging body and extended forelimbs. The tail is also used as a strut when animals are descending tree trunks or slanting branches.

EARLIER STUDIES OF HOWLERS

Carpenter's pioneering study (1934) of mantled howlers, the first long-term field study of a primate species, was carried out on Barro Colorado in the early 1930s. His field work extended over parts of two annual cycles and was aimed at compiling a general monograph of howler behavior, particularly details of social behavior, as well as making a census of all howler troops on the island. A long hiatus in howler field studies followed his work. The next howler study was done on Barro Colorado in 1951 by Collias and Southwick (1952). They carried out an island-wide census and found surprisingly few animals. It is generally agreed that an epidemic of yellow fever in the mid-1940s may have destroyed much of the population a few years prior to this study (Galindo, personal communication). Since then, a series of studies, largely short term, have been done on Barro Colorado howlers. All of these studies has been confined primarily to observations on howler behavior in the Lutz Ravine, an area adjacent to the laboratory clearing. Researchers have focused on various aspects of behavior, including general social behavior (Altmann 1959; Bernstein 1964), patterns of activity (Chivers 1969; Mittermeier 1973), comparative activity pat-

terns of howlers and spider monkeys (Richard 1970), and features
of the howler diet (Hladik and Hladik 1969; Smith 1977).

Recently, several researchers have looked at howler behavior
in areas other than Barro Colorado. Glander (1975) carried out a
long-term study of the ecology of A. *palliata* in a riparian forest
area in Guanacoste, Costa Rica; C. Jones (in prep.) and M. Clarke
(in prog.) have worked or are working on certain aspects of howler
behavior in this same study area. Coelho et al. (1976) looked at
feeding ecology of howlers (A. *villosa pigra*) for two months in
Tikal, Guatemala, and H. Schlichte (1978) and J. Cant (in prep.)
have continued observations on howlers in Tikal. A team of researchers is currently examining the ecology of A. *palliata* in Vera-
cruz, Mexico (Estrada et al.). In addition, Neville (1972) has
examined the ecology of red howlers (A. *seniculus*) in Venezuela,
and J. Eisenberg and R. Rudran are carrying out a long-term study
of howler behavior in this same area. Gaulin (in prep.) recently
completed an examination of howler feeding ecology (A. *seniculus*)
in a montane forest in Colombia. I have examined the ecology of
red howlers (A. *seniculus*) in Peru, black howlers (A. *caraya*) in
Argentina, and red-handed howlers (A. *belzebul*) and brown howlers (A. *fusca*) in Brazil. Until quite recently, however, the general
picture of howler monkeys in the literature has come almost entirely from studies done on Barro Colorado, since data from these
more recent studies are only now becoming available.

DATA COLLECTION

More than 500 hours were spent in the field during my first two and
a half months on Barro Colorado (March-May 1974), surveying the
forest, selecting two study troops, and working out details of meth-
odology. I wanted to observe two different troops living in somewhat different types of habitat, both to increase the overall sample
size and to be able to compare responses to two somewhat different
habitats.

To reduce variables, I wanted the study troops to be approximately the same size and to represent the mean size for howler
troops on Barro Colorado at the time of my study. I took a census

of eleven troops living in different areas of the island and found that the mean troop size was 17 animals (range 13 to 27).

Study Areas

As all previous studies of howler behavior on Barro Colorado had been done in an area known as the Lutz Ravine (see figure 2.2), I felt it was imperative to select one study troop from this area in order to build on previous work and perhaps to reconcile some of the inconsistencies in the data. The Lutz Ravine is largely composed of secondary forest that is estimated to be from 60 to 80 years old (Knight 1975).

My second study site was located approximately 1.5 to 2 km, SW, of the laboratory clearing in an area bordered by the junction of

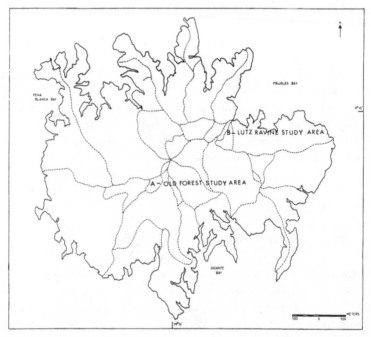

Figure 2.2.   Location of Study Areas on Barro Colorado Island, Panama Canal Zone
Area A = central portion of home range of Old Forest study troop: Area B = central portion of home range of Lutz Ravine study troop.

the Zetek and Armour trails. This area is largely composed of mature forest and appears for the most part to be undisturbed. The forest here is estimated to be at least 100 to 200 years old. For this reason, I refer to the area as the Old Forest.

Study Troops

In my initial two months' work, I found two troops that seemed to use the Lutz Ravine as a central home range area, as well as other troops that used portions of it. One troop was going through a period of social change, with immigration and emigration of males, so I chose the other troop, which appeared to be more stable in composition. This was the "Scarface" troop that had been observed in the past by Chivers (1969) and Richard (1970). When I began my study, the troop had a total of 4 adult males, 7 adult females, and 6 immature animals. Of the 11 adult animals, 8 were clearly distinguishable as individuals. Seven of them had tails that had been freeze-branded in distinctive patterns and one had natural blond spots along a portion of her tail. Thus it was relatively simple to recognize this troop when encountered.

My Old Forest study troop was exactly the same size and composition as the Lutz Ravine troop. Only one animal, however, was clearly recognizable, an adult male with natural blond spots along his tail. Initially, as this troop showed considerable overlap with neighboring troops, I had difficulty in rapidly identifying it, but as the study progressed I became more familiar with its individuals, travel patterns and home range borders and had no further problems in locating and identifying it.

Sampling Program

The behavior of each of the two study troops was sampled intensively for five days during each month of a period which included five months of the wet season of 1974 (July–November), three months of the dry season of 1975 (February–April), and the transition season of 1975-76 (mid-December to mid-January). This gave me ten sample days of howler behavior each month and five days of data on each troop. Generally I would sample the behavior of one troop for five consecutive days, spend the next several days mapping travel routes and identifying food species, and then begin my second sample on the other troop. Samples were always taken in the same order: first the Old Forest troop and then the Lutz Ravine troop.

On a typical day of intensive sampling, I would arrive in the study area well ahead of the dawn chorus, generally around 5:30 a.m. I would station myself in the area where I expected the troop to be from a prior reconnaissance and use the dawn chorus to pinpoint exact troop location. I would position myself near the troop and begin observations as soon as it was light enough to see. Observations were recorded for twelve hours each day, at five-minute intervals (see chapter 4 for details of sampling techniques).

Features Observed

In my field work, my objective was to collect data on the basic features of the howler feeding ecology: potential food sources, diet, ranging behavior, and time spent foraging. My specific expectations, methodologies, and results for each of these features are presented in the following four chapters.

3 POTENTIAL FOOD SOURCES

PREVIOUS RESEARCHERS had described Barro Colorado howlers as having abundant and omnipresent foods which they took from a limited number of common tree species, moving little to obtain such foods. This implies that the tree species used by howlers as important food sources have relatively high densities, are uniformly or randomly distributed, and provide food continuously. The last implication requires either that howlers depend mainly on perennial foods, which are continuously available in a tropical forest, or that the "limited" number of species used as food sources provide a continuous supply of seasonal items.

Given what was already known about the structure, composition, and phenology of the Barro Colorado forest, I did not expect that howler food sources would have such features. Rather, I expected that most potential food species would have relatively low densities, would tend to be clumped in their distribution, and would vary in the type and amount of food they provided over an annual cycle.

To test these different sets of expectations, I collected data on four basic features of the trees in my two study areas: species diversity; species density; spatial pattern; and temporal pattern (phenology).

METHODOLOGY

Sample Quadrats

SIZE AND LOCATION. Data on trees were collected from sample quadrats in each study area. Square quadrats 1 hectare in size were used. Three such samples were taken in each study area because of possible differences in the structure and composition of the forest within these areas. Quadrats were placed in such a way as to get the broadest coverage of the study areas, which were the presumed home-range areas of the two study troops (about 30 hectares each). By the time these samples were taken, the study troops had been under observation for several months and their home-range areas were fairly well defined (see figures 3.1 and 3.2).

With the help of two co-workers, T. Milton and A. Tarak, I laid out three 1-hectare sample quadrats in the Old Forest area. The quadrats were plotted with a 20-meter surveyor's tape and a Suunto compass. R. Thorington, A. Tarak, and R. Rudran laid out sample quadrats in my other study area, the Lutz Ravine, in the same way.

TREES SAMPLED. From past research, it was known that howlers on Barro Colorado were highly arboreal and generally fed and moved in the largest forest trees (Carpenter 1934). Thus our examination was restricted to those trees that seemed large enough to support the weight of one or more 5 to 9 kg howler and make some contribution to diet. A circumference of 60 cm breast height was selected as a cutoff point and all trees of this girth or more were included in the sample and tagged with a numbered metal tag.

MAPPING. The locations of all tagged trees within the quadrats were determined by measuring distances and angles to reference points with a surveyor's tape and Suunto compass. These locations were then plotted on maps.

SPECIES IDENTIFICATION. Many tree species in each plot could easily be identified, as most of us had prior experience working in the Barro Colorado forest. Unknown species were keyed out in the Barro Colorado herbarium, using samples removed with a tree pruner or a sketch of leaves, flowers, or fruits, as well as descriptions of bark or sap. In the Old Forest quadrats, 93 percent of all trees were identified as to genus and/or species, and in the Lutz Ravine, 98 percent.

Features Measured

DIVERSITY. Overall, diversity was measured simply by relating the total number of different species to the total number of indi-

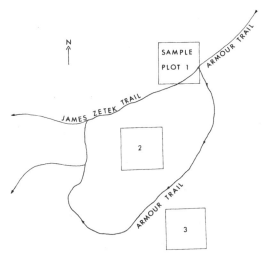

Figure 3.1.   Locations of Three Sample Plots in the Old Forest Study Area
Each plot = 1 hectare = 10,000 m². Total area sampled in the Old Forest =
30,000 m².

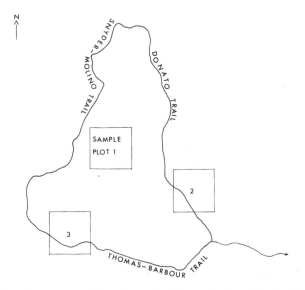

Figure 3.2.   Locations of Three Sample Plots in the Lutz Ravine Study Area
Each plot = 1 hectare = 10,000 m². Total area sampled in the Lutz Ra-
vine = 30,000 m².

vidual trees. For comparative purposes, the Shannon-Wiener function was used (Pielou 1969; Lang 1969). This function is defined as $H = -\Sigma \, p_i \log p_i$ where $p_i =$ the sampling probability. Thus, $H = -\Sigma \, (N_i/N) \log (N_i/N)$ where $N =$ the total number of individuals of all species and $N_i =$ the number of individuals of the ith species.

DENSITY.   Density was measured for each species by determining the mean number of individuals per unit area (Greig-Smith 1967; Kershaw 1973). Density was broken down arbitrarily into four classes: (i) with 10 or more individuals per hectare, (ii) with $10 >$ ii $\geqslant 5$ individuals per hectare, (iii) with $5 >$ iii $\geqslant 1$ individuals per hectare, and (iv) with less than one individual per hectare.

The relative density was measured for each species by relating the number of individuals for the species to the total number of trees and expressing the result as a percentage (Kershaw 1973).

SPATIAL PATTERN.   The purpose was to determine if the pattern of individuals of a species tended to be uniform, random, or contagious (clumped). The measure used was the ratio of the variance to the mean number of individuals per unit area (Greig-Smith 1967; Pielou 1969; Kershaw 1973). The underlying assumption is that if the pattern is random, it will be approximated by a Poisson distribution. Under a Poisson distribution, the variance is equal to the mean and therefore the expected value of the variance/mean ratio is 1. If this ratio is greater than 1, then the pattern is tending to be clumped; if the ratio is less than 1, then the pattern is tending to be uniform. The difference from the expected value can be tested for significance with a t test in which $t = (O - E)/s$ where $O =$ the observed value, $E =$ the expected value, and $s = \sqrt{2/(n - 1)}$ (Greig-Smith 1967; Kershaw 1973).

In this analysis, the results are affected by the size of the quadrat used (Greig-Smith 1967; Kershaw 1973). Therefore, three sizes of quadrat were used: 20 × 40 meters, 40 × 40 meters, and 100 × 100 meters (one hectare). This was done not only to reveal patterns that might otherwise have been missed, but also to give a better definition of pattern (i.e., to indicate the scale at which clumping might occur).

PHENOLOGY.   In August of 1974, a sampling program on tree phenology was initiated on Barro Colorado as part of an Environmental Sciences Program (ESP). Each week, Bonafacio de Leon, a member of the ESP team, spent two days walking a predetermined route through the forest, noting the phenological state of 394 canopy trees representing 145 species and forty-six families. The phenological state of each tree with respect to leaves, fruit, and flowers was recorded on a prepared data sheet. Because of the heterogenous composition of the Barro Colorado forest and the synchrony in phenology of many tree species, these data could be used to indicate island-wide conditions for any given week of the sample.

FICUS PHENOLOGY. As previous studies had indicated that fruits from species of the genus *Ficus* were important in the howler diet, I was particularly interested in the phenology of *Ficus* species (Carpenter 1934; Hladik and Hladik 1969; and others). All adult *Ficus* trees in the Lutz Ravine area (some 25 hectares) had been located, tagged, and mapped as part of a study on the feeding ecology of the fruit bat, *Artibeus jamaicensis* (Morrison 1975, 1978). The phenological state of these trees had been monitored on an individual basis every two weeks for the two-year duration of the bat project and this monitoring had then been continued by the ESP study. From this work, it was known that individual trees of this genus tended to show intraspecific asynchrony in phenology with respect to fruit production. During my study, I was given fruit production data from this project as they were collected. As no data on the phenological patterns of *Ficus* species were available from my second study area in the Old Forest, I located and tagged all adult *Ficus* trees in this area and monitored their fruit production every two weeks for the duration of my study. This gave me comparable data on fruit production by *Ficus* individuals in each of my two study areas.

RESULTS

Species Diversity

The results confirm that the Barro Colorado forest is characterized by a great diversity of species. Combined data from the six sample plots (total area: 60,000 m²) show a total of 1,017 trees ≥

TABLE 3.1. Species Diversity in Sample Plots

Sample Plots	Species	Trees	Shannon-Wiener Function
Combined	135	1,017	—
Old Forest	91	473	3.921
Lutz Ravine	100	544	3.842

60 cm in girth, representing a total of 135 different species (see table 3.1).

Sample plots in the Lutz Ravine had more individual trees per hectare than those in the Old Forest; however, the number of species per sample plot was approximately the same in both study areas (see table 3.2).

A comparison of the two study areas, using the Shannon-Wiener function, shows a diversity of 3.842 in the Lutz Ravine and 3.921 in the Old Forest. Thus, although there is more species diversity in the Old Forest, the difference is very slight.

Species Density

Combined data from the six sample plots show that 65 percent of the tree species occurred less than once per hectare; 27 percent occurred at least once but less than five times per hectare; and only 7 percent occurred five times or more per hectare (see table 3.3 and figure 3.3). Thus there were few, if any, really "common" species.

Densities of tree species between the two areas are compared in table 3.3. The Lutz Ravine has a few more species in classes I and IV, but in general the difference in density classes between the two areas is slight.

The relative density of a species indicates the probability of encountering this species within a given area. In the combined data from both study areas there is only one species, *Alseis blackiana*, with a relative density greater than 5 percent. The probability of encountering any given species in the Barro Colorado forest is very low (see table 3.4).

When the study areas are compared, the data show that there are notable differences in the relative densities of many species and

TABLE 3.2. Tree Data from Sample Quadrats

	Old Forest			Lutz Ravine		
	Quadrat			Quadrat		
	1	2	3	1	2	3
Number of trees	214	168	173	208	200	201
Number of species	58	52	57	60	63	58

Most abundant Species

Quadrat 1
Jacaranda copaia (18)
Gustavia superba (16)
Alseis blackiana (12)
Alchornea costaricensis (9)
Hura crepitans (9)
Cordia alliodora (8)
Pterocarpus rohrii (8)
Pseudobombax septenatum (6)
Hasseltia floribunda (5)
Zanthoxylum panamense (5)
Luehea seemannii (5)

Quadrat 2
Trichilia cipo (24)
Virola sebifera (9)
Alseis blackiana (9)
Quararibea asterolepis (8)
Tetragastris panamensis (6)
Heisteria concinna (6)
Hura crepitans (5)
Beilschmiedia pendula (5)
Hirtella triandra (4)

Quadrat 3
Quararibea asterolepis (16)
Poulsenia armata (13)
Virola surinamensis (12)
Guatteria dumetorum (8)
Trichilia cipo (7)
Gustavia superba (6)
Alseis blackiana (5)
Cecropia insignis (4)
Virola sebifera (4)

Quadrat 1
Macrocnemum glabrescens (22)
Hyeronima laxiflora (19)
Alseis blackiana (17)
Anacardium excelsum (11)
Luehea seemannii (10)
Poulsenia armata (9)
Ficus yoponensis (8)
Gustavia superba (7)
Pterocarpus rohrii (6)
Virola sebifera (6)

Quadrat 2
Alseis blackiana (23)
Pterocarpus rohrii (12)
Spondias radlkoferi (11)
Trophis racemosa (8)
Luehea seemannii (7)
Gustavia superba (7)
Macrocnemum glabrescens (6)
Swartzia simplex (6)
Hasseltia floribunda (5)
Spondias mombin (5)
Hyeronima laxiflora (5)

Quadrat 3
Virola surinamensis (17)
Macrocnemum glabrescens (13)
Anacardium excelsum (12)
Luehea seemannii (11)
Virola sebifera (9)
Swartzia simplex (6)
Spondias radlkofera (7)
Gustavia superba (6)
Spondias mombin (5)
Alseis blackiana (5)

Note: Above data include Palmae, which are not included in other tables and figures.

that various species are found in only one area or the other. For example, the relative density of *Jacaranda copaia* in the Lutz Ravine is 0.5 percent; in the Old Forest it is 4.2 percent. *Macrocnemum glabrescens* has a relative density of 7 percent in the Lutz Ravine, but in the Old Forest its relative density is only 0.2 percent. Overall, 41 percent of the total species from the combined data are found in both study areas.

Spatial Pattern

Because of the low densities of most tree species in the sample, there were not many species that could be tested for pattern meaningfully. It was decided to test only those species that had 9 or more individuals in the combined data (i.e., 60,000 m² of forest). There were 34 such species, and 23 of these were tested for pattern.

The results show that 20 of the 23 species were significantly clumped at at least one of the scales used (i.e., 20 × 40 m, 40 × 40 m, or 100 × 100 m = one hectare); 11 species were significantly clumped at all three scales; 4 were clumped only at the hectare scale. Only 3 had patterns that were not significantly different from random (see table 3.5).

Species that are clumped can be described as "patchy" in space, and species that have very low densities can also be described as "patchy"—if an individual tree is considered as a patch of food. Thus, at the level of particular species, most of the potential food sources for howlers can be described as patchy in space.

Phenology

Combined data covering an annual cycle (August 1974–July 1975) show that there were strong peaks and valleys in the production of seasonal items (new leaves, flowers, and fruit) by canopy trees on Barro Colorado. Production of seasonal items was highest in the late dry season and early wet season (March–June) and lowest in the late wet and transition season (October–January). This was also the pattern indicated by the data of Croat (1967, 1978) and Foster (1973). Four years of additional data, collected since my study, have confirmed that these are persistent annual patterns (ESP, in prep.; see also Leigh and Smythe, 1978).

TABLE 3.3. Species Density in Sample Plots

DENSITY: COMBINED DATA

Class	No. Individuals per Hectare	Species
I	>10	1
II	5 ≤ II < 10	9
III	1 ≤ III < 5	37
IV	<1	88
		Total 135

DENSITY: OLD FOREST

Class	No. Individuals per Hectare	Species
I	>10	1
II	5 ≤ II < 10	7
III	1 ≤ III < 5	39
IV	<1	44
		Total 91

DENSITY: LUTZ RAVINE

Class	No. Individuals per Hectare	Species
I	>10	3
II	5 ≤ II < 10	9
III	1 ≤ III < 5	32
IV	<1	56
		Total 100

Leaf production was highest at the onset of the wet season (May–June); flower production peaked in the dry season, while fruit production showed two peaks—one in the late dry to early wet season (April–May) and the other in the mid-wet season (August–September). Thus, the supply of seasonal items varied over an annual cycle (see figure 3.4).

To get an idea of how long seasonal items might be available from particular species, I drew 12 species at random from the 145 species represented in the ESP sample and noted the months in which new leaves, fruits, and flowers, respectively, were observed on these species (table 3.6). As shown in table 3.6, young leaves were avail-

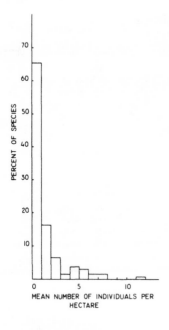

Figure 3.3. Species Density Percent of tree species having mean densities of >0 and ≤1 individual per hectare, etc., in combined data from three one-hectare sample plots in Old Forest and Lutz Ravine, in which all trees with circumference breast height ≥60 cm were counted.

TABLE 3.4. Relative Density of 47 Most Abundant Species in Combined Data from Sample Plots

Species	N	Per Hectare	Relative Density (percent)
Alseis blackiana	70	11.67	6.27
Astrocaryum standleyanum	47	7.83	4.21
Gustavia superba	46	7.67	4.12
Macrocnemum glabrescens	42	7.00	3.76
Trichilia cipo	39	6.50	3.49
Virola surinamensis	36	6.00	3.22
Luehea seemannii	34	5.67	3.04
Hyeronima laxiflora	33	5.50	2.95
Virola sebifera	32	5.33	2.86
Quararibea asterolepis	31	5.17	2.78
Poulsenia armata	29	4.83	2.60
Pterocarpus rohrii	28	4.67	2.51
Spondias radlkoferi	25	4.17	2.24
Anacardium excelsum	25	4.17	2.24
Jacaranda copaia	24	4.00	2.15
Alchornea costaricensis	19	3.17	1.70
Platypodium elegans	19	3.17	1.70

TABLE 3.4. (*Continued*)

Species	N	Per Hectare	Relative Density (percent)
Hura crepitans	17	2.83	1.43
Swartzia simplex	16	2.67	1.34
Ficus yoponensis	15	2.50	1.25
Cordia alliodora	14	2.33	1.16
Heisteria concinna	13	2.17	1.16
Trophis racemosa	13	2.17	1.07
Zanthoxylum panamense	12	2.00	1.07
Pseudobombax septenatum	12	2.00	1.07
Guatteria dumetorum	12	2.00	1.07
Scheelea zonensis	11	1.83	0.98
Ficus insipida	11	1.83	0.98
Protium tenuifolium	11	1.83	0.98
Hasseltia floribunda	11	1.83	0.98
Apeiba membranacea	10	1.67	0.90
Protium panamensis	10	1.67	0.90
Spondias mombin	10	1.67	0.90
Tetragastris panamensis	9	1.50	0.81
Sapium caudatum	8	1.33	0.72
Hirtella triandra	7	1.17	0.63
Beilschmiedia pendula	7	1.17	0.63
Maquira costaricana	7	1.17	0.63
Cavanillesia platanifolia	7	1.17	0.63
Guapira standleyanum	7	1.17	0.63
Tachigalia versicolor	7	1.17	0.63
Cupania papillosa	6	1.00	0.54
Prioria copaifera	6	1.00	0.54
Dipteryx panamensis	6	1.00	0.54
Faramea occidentalis	6	1.00	0.54
Inga goldmanii	6	1.00	0.54
Cecropia insignis	6	1.00	0.54

able on particular species for a mean of 6.81 months of an annual cycle, green and ripe fruits for 3.67 months, and flower buds and flowers for 2.73 months. Ripe fruits were available on particular species for only 1.13 months. Further, these items were available on individual trees for even shorter periods: new leaves for a mean of 5.26 months, green and ripe fruits for 2.08 months, and flower buds and flowers for 1.84 months. Ripe fruits were available on individual trees for only 0.78 months.

TABLE 3.5. Spatial Pattern of Tree Species

Species ( ) = number of trees	20m × 40m n = 60		40m × 40m n = 24		100m × 100m n = 6		Conclusion
	s^2/x	p	s^2/x	p	s^2/x	p	
Hyeronima laxiflora (31)	2.14	.005	4.17	.005	9.71	.005	Significantly clumped at all three scales
Anacardium excelsum (25)	2.50	.005	3.18	.005	8.10	.005	
Quaraibea asterolepis (30)	1.59	.005	2.16	.005	7.92	.005	
Poulsenia armata (28)	1.73	.005	2.12	.005	6.06	.005	
Pterocarpus rohrii (28)	1.38	.025	2.37	.005	5.63	.005	
Trichilia cipo (36)	2.18	.005	3.46	.005	14.33	.005	
Macrocnemum glabrescens (43)	3.54	.005	3.80	.005	11.91	.005	
Jacaranda copaia (24)	2.31	.005	4.00	.005	12.10	.005	
Virola surinamensis (35)	1.59	.005	1.51	.05	7.25	.005	
Alseis blackiana (68)	1.71	.005	2.61	.005	4.54	.005	
Luehea seemannii (32)	1.44	.01	2.05	.005	3.20	.01	
Gustavia superba (45)	1.53	.005	1.24	.10	2.49	.05	Significantly clumped at at least one scale
Spondias mombin (11)	1.25	.10	1.74	.01	3.36	.01	
Spondias radlkoferi (25)	1.13	.10	0.82	.10	4.16	.005	
Cordia alliodora (16)	1.20	.10	1.35	.10	2.65	.025	
Hura crepitans (16)	1.09	.10	1.74	.01	5.05	.005	
Pseudobombax septenatum (10)	1.09	.10	1.04	.10	3.52	.01	
Ficus yoponensis (14)	0.98	.10	1.51	.05	2.86	.025	
Virola sebifera (32)	1.09	.10	2.04	.005	2.00	.10	
Platypodium elegans (11)	1.32	.05	1.22	.10	0.75	.10	Not significantly different from random
Ficus insipida (11)	0.86	.10	0.74	.10	1.40	.10	
Zanthoxylum panamense (9)	0.86	.10	0.96	.10	1.53	.10	
Protium panamensis (10)	1.09	.10	0.74	.10	0.64	.10	

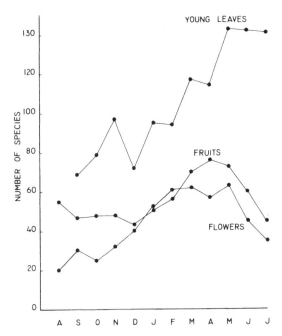

Figure 3.4.   Availability of Seasonal Foods
Number of tree species on which young leaves, fruits, and flowers were
observed in weekly phenological monitoring program, covering 145 species.

Data also show that over an annual cycle the mean number of species per month with mature leaves was 143 compared with a mean of 103 with some new leaves, 56 with fruit, and 44 with flowers. In the random sample of 12 species, mature leaves were available on particular species for a mean of 11.75 months of an annual cycle, and on individual trees for 11.03 months. This supports one's impression that mature leaves are far more abundant and far more often available than seasonal items in the Barro Colorado forest.

A further examination of the data reveals that most species represented in the sample by more than one individual showed at least some degree of intraspecific synchrony in phenology. In some species, such as *Hirtella triandra*, the synchrony was relatively tight; in others, such as *Ceiba pentandra*, it was relatively loose. Except in a few cases in which the synchrony was perfect,

TABLE 3.6. Availability of Potential Foods from Twelve Tree Species
Drawn at Random

A	Number of Months Seasonal Foods Available		
Species	Young leaves	Fruits	Flowers
Hirtella triandra	10	4	4
Jacaranda copaia	7	5	2
Tabebuia rosea	10	2	6
Ceiba pentandra	12	2	2
Apeiba membranacea	9	11	7
Triplaris cumingiana	9	2	4
Palicourea guianensis	11	6	3
Ormosia sp.	8	0	2
Ocotea cernua	6	3	3
Paullinia turbacensis	4	7	3
Zuelania guidonia	4	3	3
Cecropia insignis	8	8	6

B	Mean Number of Months in which Items Were Observed on Twelve Tree Species Drawn at Random	
Food Category	Species	Individual Trees
Young leaves	6.81 ± 2.53	5.26 ± 2.53
Mature leaves	11.75 ± 0.46	11.03 ± 1.22
Green and ripe fruits	3.67 ± 2.92	2.08 ± 1.80
Ripe fruits	1.13 ± 1.27	0.78 ± 1.00
Flower buds and flowers	2.73 ± 2.00	1.84 ± 1.12

the periods during which *most* individuals of a species had new
leaves, fruits, or flowers were much shorter than the periods during which one or some individuals had such items.

Most of these seasonal items also appeared to be very ephemeral
in time in terms of edibility. Based on qualitative observations, I
would estimate that for howlers most seasonal food items are optimally edible for much shorter periods of time than they are actually present on the tree. On individual trees, my subjective impression was that most synchronously produced young leaves were
optimally edible for only some two to four days, individual ripe
fruits for three to six days, and individual flowers for one to three
days.

Thus data indicate that seasonal items from particular species and individual trees are very patchy in time, especially if edibility is considered.

Ficus Phenology

Lutz Ravine data covering a two-year period (March 4 to November 14, 1973, collected by Morrison; December 2, 1974 to February 22, 1975, collected by M. Estribi, ESP), and my data from the Old Forest study area, collected from July 1974 to May 1976, showed a clear pattern of intraspecifically asynchronous fruit production for the two *Ficus* species, *F. yoponensis* and *F. insipida*, making up the majority of the sample. There were too few individuals of other *Ficus* species in the sample to clearly discern their patterns of fruit production. Though individuals of both *F. yoponensis* and *F. insipida* were intraspecifically asynchronous, there were seasonal peaks and valleys in the overall production of fruit by the two species. Morrison (1978) found a significant correlation between two sample years of Lutz Ravine data in the number of trees fruiting in a given month ($r = .675$, $p < .05$). The peaks occurred in the periods April–June and December–January. A consistent low was also found for fruit production in August–October. I found these same patterns in the Old Forest data (figures 3.5 and 3.6). In the Lutz Ravine, however, there was no sample month when some fig fruit was not available, whereas in the Old Forest there were several months when no fig fruit was available. This presumably reflects the great difference in the number of fig trees between the two study areas (142 in the Lutz Ravine vs. 32 in the Old Forest).

When production of figs is compared with overall production of fruit by Barro Colorado tree species, it can be seen that the times of peak fruit production by *Ficus* species are out of synchrony with fruit production by many other tree species, particularly those species producing a rich, sugary pulp or aril (e.g., *Tetragastris panamensis, Spondias mombin, Brosimum alicastrum, Anacardium excelsum, Virola surinamensis*). A variety of factors might influence the timing of *Ficus* fruit crops, including factors related to pollination, competition for seed dispersal agents, competition for ger-

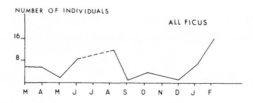

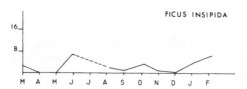

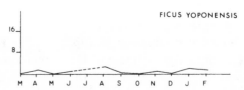

Figure 3.5.   Patterns of Fruit Production by Ficus *Trees in the Old Forest Study Area*
Data collected bi-weekly at intermittent intervals from July 1974 to May 1976. Points represent means when monthly data for more than one year were available.

mination sites, predation on seeds or fruits, and/or climactic or microhabitat influences. Whatever the reasons, the asynchronous production of fig fruit by individual trees (each one of which is estimated to produce some 40,000 fruits per crop; Hladik and Hladik 1969) and the peaks in fig production at times when other fruit production is depressed, would appear to make species of *Ficus* a potentially useful food source for howler monkeys.

Ficus insipida and *F. yoponensis* were by far the most common *Ficus* species in both study areas and accounted for 84 percent of

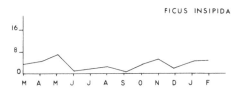

Figure 3.6. Patterns of Fruit Production by Ficus *Trees in the Lutz Ravine Study Area*
Data collected bi-weekly from March 1973 to February 1975. Points represent means for two years. From Morrison (1975).

the phenological sample for the genus in the Lutz Ravine and 63 percent in the Old Forest. In the Lutz Ravine, fruit production by *F. insipida* was highest in November and from January to May. *Ficus yoponensis* showed strong peaks in June and December. In the Old Forest, *F. insipida* showed a strong peak in fruit production in February (two years data) and another in June (one year data). Old Forest data on *F. yoponensis* do not show notable peaks for any month, but since there were only five trees of this species in the area, data are inconclusive. Further, data from the Lutz Ravine indicate that fruit production of the two fig species peaks at

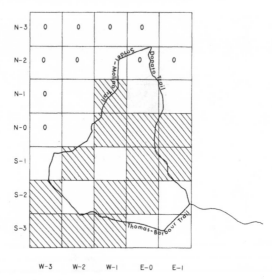

Figure 3.7.   Grid of Central Portion of Lutz Ravine Study Area
The distribution of individuals of two Ficus species, F. insipida *and* F.
yoponensis *are shown. Each quadrat = 1 hectare. Shaded quadrats had
five or more individuals of the two species. Quadrats with "0" in them had
no individuals of either species.*

somewhat different times, which would reduce competition between them for dispersal agents.

In both study areas, many trees of these two species tended to
produce more than one fruit crop per annum. In the Lutz Ravine,
F. yoponensis produced 1.13 (±.61) crops per annum and *F. insipida*
.93 (±.46) crops per annum, while in the Old Forest, *F. yoponensis*
produced 1.6 (±.76) crops per annum and *F. insipida* 1.0 (±.89)
crops per annum. Thus individual *F. yoponensis* showed a tendency
to produce more fruit crops per annum than *F. insipida* (*t*-test, two-
tailed test, $p < .005$ for Lutz Ravine (Morrison, 1975), $p < .05$
for Old Forest).

Two and a half additional years of phenological data have since
been collected and analyzed for the 105 individuals of *F. yoponen-*
sis and *F. insipida* found in the Lutz Ravine. These data, which
cover four and a half years of continuous observation, confirm that
the trends discussed above are persistent characteristics of these

two *Ficus* species (Milton et al, in press). Individuals of both species produce fruit at different times each year by initiating fruit crops at intervals of less than a year but more than a half year. For *F. yoponensis* individuals, the median fruiting interval is around thirty weeks while for *F. insipida* it is around forty weeks. Since each tree is constantly altering its time of fruit production, at least some individuals of each species are likely to be producing fruit in any month of the year. This staggered system of fruit production is viewed as a type of "bet hedging" with respect to an individual tree's overall reproductive success. In both *Ficus* species, conditions optimal for pollination, seed dispersal, and germination usually do not occur concurrently. For example, during fruiting peaks such as the early to mid-dry season when pollination agents are most abundant, intraspecific competition for dispersal agents and germination sites is most intense; during times when few other individuals are fruiting, dispersal agents may be more abundant but pollination probabilities are lower. The unusual fruiting pattern of these species appears to be a device whereby each tree, over its estimated lifetime of 60 to 100+ years, is able to maximize pollination success, seed dispersal success, or both in some years by constantly altering its time of fruit production.

SUMMARY

I collected data on canopy trees on Barro Colorado to test certain implied assumptions by previous researchers regarding food sources of howler monkeys—i.e., that species used as food sources have relatively high densities, are uniformly or randomly distributed, and provide food continuously.

As shown by the data presented above, the canopy trees on Barro Colorado have the following features:

a great diversity of species
a very low density for most species
a clumped distribution (for species occurring in sufficient numbers to be tested)
pronounced peaks and valleys in the production of flush leaves, fruit, and flowers

prolonged discontinuities in the production of seasonal items by particular species

very short periods of availability of seasonal items on individual trees, especially if edibility is considered.

Thus, potential seasonal foods for howlers on Barro Colorado can be described as very patchy in both space and time.

4 THE HOWLER DIET

GIVEN THE lack of digestive specializations comparable to those of many Old World leaf-eating primates and given the problems of plant parts as food (i.e., bulkiness, variable nutritional quality, incomplete nutrients, and secondary compounds), I expected howlers to be very selective in their choice of diet, as predicted by Westoby (1974) and Freeland and Janzen (1974) for generalist herbivores. Specifically, I expected howlers: to select foods with a relatively high nutritional quality (i.e., seasonal items); to select foods so as to get a balance of nutrients; and to select foods so as to minimize the effects of compounds that reduce digestibility, such as tannins and/or potentially harmful toxins.

Being selective in feeding, however, would increase the costs of food procurement, since it would increase the amount of time spent searching for food (i.e., traveling). But if the efficiency of foraging is maximized by natural selection, as postulated by the theory of foraging strategy (Schoener 1971), then howlers should have behavioral and morphological features that would minimize the costs of procuring their preferred foods. I therefore expected them to have features that would minimize the costs of procuring seasonal items which, as shown in chapter 3, are patchily distributed in both space and time in the Barro Colorado forest. Specifically, I expected howlers: to diversify food sources; to change food sources over time; to exploit energy-rich foods intensively when such foods were available; to sample and monitor potential food sources continuously; and to specialize, to some extent, on species that might be continuous sources of preferred foods.

Thus, I expected the howler diet to reflect a compromise between the pressure to be selective with respect to food quality and the pressure to minimize the costs of foraging.

To test my expectations, I collected data to determine: the proportions of seasonal and perennial foods in the howler diet; the proportions of leaves, fruit, and flowers; the number of different species used as food sources; the rate of turnover of the species used as food sources; the relative use of different species; the preferences shown for certain species; the role of species of *Ficus* in the howler diet; and the role of certain plant families in the howler diet.

METHODOLOGY

Data on feeding behavior were collected in terms of time spent feeding. Time was used as a unit of measurement for several reasons. First, though measures of wet weight (e.g., Hladik and Hladik 1969; C. M. Hladik 1978a, 1978b) might be better for estimating the actual quantity of food eaten during a short period of observation, feeding time is a good indicator of the relative importance of food categories and food species over a long period of study; and if actual quantities of food are required for the problem being investigated, these can then be estimated with additional data on feeding rates and the wet weights of the respective food items (Chivers 1974). Second, the measure of feeding time is the one used by most other researchers who have done or are doing similar studies (Sussman 1973; Glander 1975; Clutton-Brock 1975; Rudran 1978; Terborgh and Janson in prep.); so the data collected in this study can be used for comparative purposes. Third, time is used to measure efficiency in all of the mathematical models of foraging strategy referred to earlier (e.g., MacArthur and Pianka 1966; Emlen 1966; Schoener 1971; and others).

For my observations, since the tree heights and the dense foliage made the focal animal technique impractical, I used the scan sampling technique described by J. Altmann (1974). This technique works well for howlers as they tend to perform most activities together. From dawn until dusk on each sample day, I observed the study troop at five-minute intervals and recorded the activity of each animal visible to me. Each of these animal activity states was counted as an activity record for the corresponding five-minute interval. Percentages for each activity—feeding, traveling, resting, and moving—were then calculated in relation to the total activity records for each five-minute interval. These percentages were added and divided by the total number of five-minute intervals in the sample day to arrive at the percent of time spent at each activity. Only full sample days (twelve hours of observation) were used for this purpose—forty days in the Old Forest and forty-five in the Lutz Ravine.

Feeding activity included the inspection of food, bringing it to the mouth, chewing and swallowing. Whenever an animal was feeding, the food category (e.g., flush leaf, fruit, etc.) and food species were recorded. To calculate the percent of feeding time spent on different food categories and species, I related the feeding activity records for each food category and species to the corresponding activity records for the five-minute intervals and proceeded as above; the resulting percentages were divided by the percent of time spent feeding in the sample days.

When a food species could not be identified in the field, a sample was collected or a description taken that would later allow me to key it out in the Barro Colorado herbarium. Most food items were ultimately keyed to the species level; a few, particularly species of the Leguminosae and Bignoniaceae, could only be keyed as to family and/or genus due to the great similarity of the leaves of various species of the family and a lack of flowering parts (e.g., species of *Inga*, *Ormosia* and the like). Food species identified accounted for 93 percent of total feeding time in the Lutz Ravine and 87 percent in the Old Forest.

When howlers were feeding, fruits and flowers could be readily identified as categories. Leaves were also readily identified as a category, but their relative maturity was sometimes difficult to assess. It is obvious that some subjectivity is involved in determining the relative maturity of a leaf, especially when one is viewing leaf clusters through binoculars from a distance of some 30 m. Deciduous trees that had dropped all old foliage and were putting out a new leaf crop posed no problem since all leaves were young. Further, young leaves were often rolled up and/or were a different color than mature leaves on the same tree. By studying the location of the leaves being eaten (the youngest leaves are generally at the tips of twigs or petioles) as well as examining dropped food to see which areas had been eaten and which had not been eaten and noting their respective textures, I was usually able to determine relative leaf maturity.

In the Lutz Ravine, all nine monthly samples were taken on the same study troop. In the Old Forest, because of initial problems with troop identification, two monthly samples were taken on one troop and seven on another. All nine samples in the Old Forest should be equally representative of howler feeding behavior in this area and thus are included in these analyses.

The data for the sample days from both study areas were combined to get overall results. Data from the two study areas, separately, and from the three seasons—wet, dry, and transition—were then compared for similarities and differences. In comparisons of the two areas, Mann-Whitney tests were used to determine if the differences were statistically significant. In comparisons of the three seasons, Kruskal-Wallis tests were used for this purpose. Differences were considered to be significant at levels of .05 or less.

The data on feeding behavior were related to the data on food sources to see if there were correlations. Spearman rank correlation tests were used to determine if such relationships were statistically significant. In a few cases, phi coefficient tests were used. The relationships were considered to be significant at levels of .05 or less.

The principal references for the above tests were Conover (1971) and Huntsberger and Billingsley (1973). Most of the tests were run on a CDC 6600 computer using the SPSS-Statistical Package for the Social Sciences, Version 6, April 1, 1975. Unless otherwise indicated, the p values shown for Mann-Whitney and Spearman rank tests are for two-tailed tests. Where appropriate, the results of these tests were corrected for ties.

RESULTS

Seasonal vs. Perennial Resources

OVERALL.   Combined data from both study areas show that during the sampling period 91.4 percent of overall feeding time was spent eating seasonal foods—young leaves, fruits, and flowers. Only 1.6 percent of overall feeding time was spent eating perennial foods—i.e., foliage that could be considered mature; 7 percent of overall feeding time was spent on foliage that could not be classified as to relative maturity. Thus, as expected, howlers spent an overwhelming percent of feeding time on seasonal rather than perennial foods (see table 4.1).

As noted in chapter 3, perennial foods (e.g., mature leaves) are far more abundant than seasonal foods in the Barro Colorado forest and are also always available, whereas seasonal foods are very patchy in both space and time. The lack of relationship between feeding time and the supply of these two types of food suggests that factors other than abundance and distribution are influencing the monkeys in their choice of diet—i.e., they are showing food "preferences." Thus, in this sense, there is strong evidence that howlers prefer seasonal foods. An animal may be said to "prefer" a food if the relative frequency of the food in the diet exceeds the relative frequency of the food in the habitat (Reichman 1975).

SPATIAL DIFFERENCES. In the Old Forest study area, 86.1 percent of overall feeding time was spent on seasonal food items, 2.9 percent on mature foliage, and 11 percent on leaves of uncertain maturity. Many of the leaves which could not be classified with regard to relative maturity were vine leaves.

In the Lutz Ravine a very similar pattern was noted—96 percent of overall feeding time here was spent on seasonal items, 0.5 percent on mature leaves, and 3.5 percent on leaves of uncertain maturity. But howlers in the Lutz Ravine spent significantly more time

TABLE 4.1. Seasonal vs. Perennial Foods

	Mean Daily Percent of Feeding Time			
	Wet	Transition	Dry	Overall
Seasonal Foods				
Old Forest	84.72	70.70	91.53	86.10
Lutz Ravine	95.26	90.80	98.63	96.11
Combined data	89.99	80.55	95.08	91.40
Perennial Foods				
Old Forest	2.94	7.00	1.60	2.95
Lutz Ravine	0.88	0	0.03	0.50
Combined data	1.80	3.50	0.82	1.65

eating seasonal foods than howlers in the Old Forest (M–W, p = .006, n = m = 9).

TEMPORAL DIFFERENCES. Combined as well as separate data from the two study areas show that the percent of time spent on seasonal foods was highest in the dry season, next in the wet and lowest in the transition. These differences are statistically significant (K–W, p = .043, N = 18) and correlate with seasonal differences in overall production of flush, fruit, and flowers. On a monthly basis, there is a strong correlation between time spent eating seasonal foods and overall production of such foods (r_s = 0.82, p = .015, n = 7, one-tailed test). Thus, the data suggest that howlers tend to eat a greater proportion of seasonal foods when there is a greater supply of them.

Data showed that there were some monthly samples in which one or both troops ate no leaves that could be identified as mature. In the Old Forest, the months with the highest feeding time on mature leaves were August and October (wet season) and December (transition season). Phenological data on overall flush production by canopy trees on Barro Colorado (see figure 3.4) show that December is the lowest month of the year for flush leaf production. In the Lutz Ravine, the months with the highest feeding time on mature leaves were September and November, which are also months of relatively low flush production. These relationships suggest that the monkeys eat more mature leaves when there is a lower supply of young leaves.

Food Categories

OVERALL. Combined data from both study areas show that howlers spent 48.2 percent of overall feeding time on leaves, 42.1 percent on fruit, and 9.6 percent on flowers (see table 4.2). Insects, particularly beetle larvae (Curculionidae), were inadvertently eaten with some foods (e.g., *Ficus* fruits, fruit of *Quararibea asterolepis*) and could make some nutritional contribution to the howler diet.

Leaves: Leaves are a basic part of the howler diet and were eaten on every sample day in both study areas. Further, a substantial proportion of feeding time each day was usually spent on leaf-eating.

As noted above, young leaves were overwhelmingly preferred to

TABLE 4.2. Food Categories as Percent of Feeding Time

	Mean Daily			
	Wet	Transition	Dry	Overall
Combined Data				
Leaves	48.8	84.5	35.2	48.2 ± 26.3
Fruit	46.1	9.7	46.9	42.1 ± 26.1
Flowers	4.9	5.6	17.9	9.6 ± 13.3
Old Forest				
Leaves	49.7	92.6	45.2	53.4 ± 29.6
Fruit	42.5	4.2	40.3	36.9 ± 30.8
Flowers	7.3	2.8	14.3	9.3 ± 11.5
Lutz Ravine				
Leaves	48.1	76.4	25.1	43.6 ± 22.3
Fruit	48.9	15.2	53.4	46.7 ± 20.3
Flowers	3.0	8.4	21.4	9.6 ± 14.9

mature leaves. The term "young leaf" includes all types of immature foliage, i.e., leaf buds, rolled leaflets, partially opened young shoots, and fully opened but soft and tender young leaves. Of the leaves identified, 93 percent were taken from canopy tree species; the remaining percentage were taken from various vines, lianas, epiphytes, and hemi-epiphytes, typically found growing in the branches of canopy trees.

Howlers selected young leaves of many different types, sizes, and textures. In their pursuit of immature foliage, howlers seemed little deterred by flush densely coated with hairs (*Inga goldmanii*), by extremely sharp thorns (*Poulsenia armata*), by soft thorns on the leaves and sharp thorns on the branches (*Zanthoxylum* spp.), by biting ants (*Cecropia insignis*), by presumably caustic saps (*Omphalea diandra*), or by sticky exhudations (*Ficus* spp.; *Tabernaemontana arborea*).

Fruit: Fruit was also a basic part of the howler diet and was eaten consistently in both study areas throughout the study. It was eaten on every sample day but one in the Lutz Ravine and on every sample day but eight in the Old Forest. It usually accounted for a substantial proportion of daily feeding time.

Most fruit was taken from large canopy tree species (e.g., *Ficus insipida, F. yoponensis, Brosimum alicastrum, Spondias mombin*). Howlers typically ate soft-pulped fruits and arils but on occasion ate harder fruits (e.g., immature fruits of *Cecropia insignis* or *Ficus* spp.) or selected softer parts of very hard fruits (e.g., the pericarp and mesocarp of *Dipteryx panamensis*, the mesocarp of *Socretea durissima*).

Seeds from some fruits were swallowed while others were rejected. Large seeds (e.g., *Spondias mombin*, $\bar{x}$ seed size = 1.9 cm length $\times$ 1.7 cm width, N = 5) were swallowed as well as medium-sized (e.g., *Doliocarpus* spp.) and small seeds (e.g., *Ficus* spp., *Hyeronima laxiflora*). Many seeds which were swallowed are presumed to have nutrients adhering to them that are most efficiently removed by the action of digestive enzymes (e.g., *Lacmellea panamensis*, *Hyeronima laxiflora*) (Hladik and Hladik 1969; McKey 1975). Small seeds are probably swallowed in many cases because they do no toxic damage and it is too much trouble to pick or spit them out (e.g., *Ficus* spp.). Some other seeds (e.g., *Brosimum alicastrum*) are probably rejected because of toxic chemicals that would be released into the monkey's system if too many seeds were chewed up or broken down by digestive enzymes (Janzen 1969, 1971; McKey 1975).

Flowers: Flowers and flower buds play a much smaller role in the howler diet than young leaves or fruit. They were not eaten on almost half of the sample days and during some whole sample months.

Flowers from tree species and species of vines, lianas, and epiphytes were eaten. During the dry season, when flower-eating was most pronounced, howlers ate flowers from many vine species, particularly species of Bignoniaceae. Tree and vine species of Leguminosae were another important flower source.

DAILY DIETARY MIX.   As a general rule, howlers in both study areas tended to eat both leaves and fruit each day (figure 4.1). There were only nine days in the total sample when they did not eat both food categories. All of these were days when no fruit was eaten, and seven of them were at times of lowest overall fruit production. Thus, the lack of fruit in the diet on these particular days was apparently due to a lack of available fruit sources.

On the other hand, there were days when there seemed to be enough fruit in a study area so that the troop could have spent all of

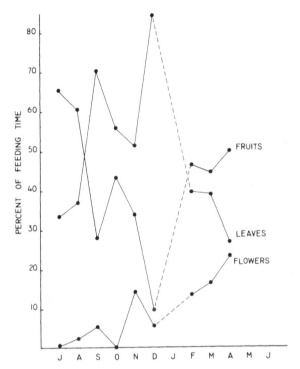

Figure 4.1.   Food Categories
Percent of feeding time spent on leaves, fruit, and flowers during monthly
samples in combined data from both study areas. Old Forest s.d. for leaves
±29.6, *fruit* ±30.8, *flowers* ±11.5. *Lutz Ravine s.d. for leaves* ±22.3, *fruit*
±20.3, *flowers* ±14.9.

its feeding time on fruit. On various occasions one or the other of
my study troops would vacate a tree in which it had been feeding
heavily and move away to feed on leaves. In this same day, one or
more other howler troops would then move into the *same* tree va-
cated by the study troop and feed heavily from it. This suggests
that factors other than availability are influencing howlers in their
choice of food categories and, in this sense, they seem to prefer to
eat both leaves and fruit each day. Presumably, foods from these
two categories together provide them with the best nutrient mix.

SPATIAL DIFFERENCES. Howlers in the Old Forest spent more

feeding time on leaves and less on fruit than howlers in the Lutz Ravine, but the differences were not statistically significant (M-W, $p = .113$, $p = .064$, $n = 40$, $m = 45$). Also, in the Old Forest, there was considerably more fluctuation over an annual cycle in the percent of time spent feeding daily on leaves and fruits, as shown by the greater standard deviations (table 4.2 and figure 4.1). Thus the proportions of leaves and fruits in the daily diet were more regular in the Lutz Ravine.

These differences between the two areas seem to be related to the above-mentioned differences in forest composition between the two study areas, particularly the greater number of fig trees in the Lutz Ravine.

TEMPORAL DIFFERENCES. In the combined data from the two study areas there were significant seasonal differences in the time spent feeding on leaves, fruit, and flowers (K-W: $p < .0001$; $p = .0003$; $p < .0001$; $N = 85$). Leaves were eaten most in the transition season, second in the wet, and least in the dry. Fruit was eaten about equally in the wet and dry seasons, but less in the transition. Flowers were eaten most in the dry season, second in the transition, and least in the wet. When analyzed separately, the data from both study areas show these same differences.

Some possible explanations for these patterns were suggested when the data on food categories were examined in relation to the data presented in chapter 3 on the production of leaves, fruits, and flowers by canopy trees.

Leaves. The transition season, when leaves were eaten most by both study troops, is the time of lowest production of flush leaves. It is also, however, the time of lowest production of fruit. Similarly, leaves were eaten more in the late wet season than in the dry, even though production of flush is lower at that time; but production of fruit is also lower. On a monthly basis, there was a strong negative correlation between time spent eating leaves and the number of species producing flush ($r_s = -.89$, $p = .01$, $n = 7$); but there was a stronger negative correlation between time spent eating leaves and the number of species bearing fruit ($r_s = -1.00$, $p < .002$, $n = 7$). Thus it would appear that howlers eat more leaves when there is less fruit available.

Leaves are the main source of protein for howlers whereas fruit is their main source of ready energy (see Hladik et al. 1971; Smith 1977; Milton 1979a). But leaves contain a small amount of ready energy (some 3–4 percent dry weight) as well as large amounts of cell wall material, some of which might, through the end products of fermentation, contribute energy to howlers. As leaves—even young leaves—are less patchy than fruit in tropical forests, they can generally be found with less search and therefore less expenditure of energy. On the other hand, if fruit is found, the reward for such an expenditure is a concentrated source of ready energy. When fruit is less available, the expenditure of energy in search is less likely to be rewarded, and there may be times when it is more profitable on a net basis for howlers to eat leaves for energy as well as protein than to search for fruit. The proportion of leaves in the howler diet, at least above a certain minimum required for protein, may thus depend on the availability of fruit (and/or flowers). There were five days in the sample period when 100 percent of feeding time was spent on leaves, all but one in the period of lowest fruit production. During these days, howlers were apparently relying on leaves for all of their nutritional needs, including energy. Given the low ready energy content of leaves, the limitation in energy returns to howlers through fermentation (Milton et al., in press) and the low protein content of many leaf species, howlers may face real nutritional problems when living on diets high in foliage (see, for example, Nagy and Milton 1979a, 1979b). In such cases, the animals' fat reserves (if any) may be used to help supply the energy necessary for daily maintenance activities.

Fruit: On a monthly basis, there was a strong positive correlation between time spent eating fruit and the number of species bearing fruit ($r_s = .93$, $p < .01$, $n = 7$). This indicates that howlers eat more fruit, up to a point, when there is more fruit available. The upper limits to fruit eating may be set by demands for nutrients best supplied by leaves (i.e., protein). Howlers typically eat fruit in the morning but at some point during the day they apparently must eat leaves, which they typically do in the late afternoon.

During the transition season, when fruit production is greatly depressed, the time spent eating fruit dropped to almost zero, particularly in the Old Forest. Since fruit is the best source of ready energy for howlers, they must have had difficulty at this time in meeting their energy requirements. Presumably if fruit had been available, howlers would have eaten some. A thorough examination of the forest during this sample period confirmed that there was indeed little or no fruit available. Even sources of young leaves seemed limited, as evidenced by the fact that three troops were competing for one leaf source (*Ficus insipida* flush) in the Lutz Ravine and two troops were competing for two leaf sources (*Ceiba pentandra* flush; *Cecropia insignis* flush and petioles) in the Old Forest.

Flowers: Flower eating by howler monkeys seems closely related to flower availability. On a monthly basis there is a strong positive correlation between flower eating and the number of species flowering (r_s = .79, p = .038, n = 7). Of the three food categories, flowers are by far the patchiest resource. Not only does the overall phenological patterning of flower production show a strong dry season peak, but flowers of many species are extremely ephemeral in nature, some flowers living no more than twenty-four hours.

Chemical analyses carried out on the nutritional components of some flower species from the Barro Colorado forest show that certain flowers are very rich in protein (e.g., male inflorescences of *Gnetum leyboldii*) while others are low (e.g., *Dioclea reflexa*); some flowers are also quite high in nonstructural carbohydrates (e.g., *Tabebuia rosea*). Thus many flowers are quite nutritious; they often have a protein content as high as young leaves, a nonstructural carbohydrate content as high as some fruits and a relatively low structural carbohydrate content (see pages 84–85, see also, C. M. Hladik 1978 a and b). Therefore, flowers might be able to fill various dietary needs more completely than either leaves or fruit alone. During the dry season, although there was a relatively large supply of young leaves and fruit available, howlers increased their consumption of flowers. This supports the view that many flowers are nutritious dietary items for howler monkeys and a food category that they prefer to exploit even when flush leaves and fruit are also relatively abundant.

On a daily basis there is a significant negative correlation between the time spent eating flowers and the time spent eating fruit $(r_s = -.27, p = .012, n = 85)$. This suggests that flowers substitute for fruit in the diet and are used primarily as a source of nonstructural carbohydrates. There is also a negative correlation between time spent feeding on flowers and time spent on leaves but it is less strong $(r_s = -.13, p = .22, n = 85)$. This suggests that at times flowers also serve as a source of protein.

Number of Species Eaten by Howlers

OVERALL.   In the combined data from both study areas, howlers ate a total of 109 different species during the sample period (see table 4.3). These 109 identified species accounted for 90 percent of total feeding time: 87 were used as leaf sources, 36 as fruit sources, and 25 as flower sources. Thus the number of species used as leaf sources is more than double the number of fruit species and more than triple the number of flower species. As table 4.8 shows, a few of these species were used for foods in all three dietary categories, some for foods in two and most for foods in only one.

When the species eaten by howlers from times other than the sample period as well as those noted by previous researchers on Barro Colorado (Carpenter 1934; Hladik and Hladik 1969; Mittermeier 1973) are added to the 109, the total number of species eaten is at least 125, and this is a conservative estimate.

SPATIAL DIFFERENCES.   When examined separately, both study areas showed the same pattern in the number of food species used during the sample period. The total number of species eaten was exactly the same. Also, as shown in table 4.3, the number of species used in each of the three dietary categories is strikingly similar. Thus, despite the differences in the forest between the two areas,

TABLE 4.3. Number of Different Species Eaten

	Old Forest	Lutz Ravine	In Common	Total Different
Total	73	73	37	109
Leaves	59	59	31	87
Fruit	25	23	12	36
Flowers	13	16	4	25

both troops diversified the species used as food sources to the same extent, at least as measured simply by the number of species used.

Though the same number of species were eaten in each study area, these were not necessarily the same species. In fact, only 37 species were eaten in both areas. Of the top 33 food species in each area, 16 were eaten in one area and not in the other during the sample period. Of these 16 species, 10 had higher relative densities in sample plots of the area where they were eaten than in the area where they were not eaten. Only 3 had higher relative densities in the area where they were not eaten. The association between whether or not a species was eaten and its relative density was found to be statistically significant (phi coefficient $= .50$, $p = .0358$, $n = 13$). Thus, the differences in the species used as food sources by the two study troops appear to be mainly the result of differences between the two habitats.

DAILY PATTERN. In the combined data from both study areas, howlers ate a mean number of 7.66 different species daily (see table 4.4). They ate a mean of 5.11 leaf species, 1.74 fruit species, and 0.81 flower species. There is a positive correlation between the number of species eaten daily and the percent of time spent eating leaves ($r_s = .293$, $p = .007$, $n = 85$). On the other hand, there is a negative correlation between the number of food species eaten daily and the percent of time spent eating fruit ($r_s = -.413$, $p = .001$, $n = 85$).

These correlations suggest that the pressures on howlers to diversify food species are stronger when they are eating leaves than when they are eating fruit. This may be because young leaves are less patchy in space than fruit—i.e., they are less concentrated, less discontinuous. Often during the sample period my study troop would split up and eat leaves from two or more trees of different species at the same time ("scatterfeed"), apparently because there were not enough young leaves on one tree for the whole troop; however, they rarely did this while eating fruit. Leaf species may also be diversified more on a daily basis because leaves have higher contents of potentially harmful secondary compounds. While eating leaves, howlers may have to mix species to avoid ingesting too much of any one secondary compound, as Freeland and Janzen

TABLE 4.4. Number of Different Food Species

	Mean Daily			
	Wet	Transition	Dry	Overall
Combined Data				
Leaves	5.16	6.80	4.47	5.11
Fruit	1.60	0.60	2.33	1.74
Flowers	0.40	0.90	1.40	0.81
All categories	7.16	8.30	8.13	7.66
Old Forest				
Leaves	5.40	6.80	5.07	5.45
Fruit	1.15	0.40	2.53	1.57
Flowers	0.40	0.60	1.80	0.95
All categories	6.95	7.80	9.40	7.97
Lutz Ravine				
Leaves	4.96	6.80	3.87	4.80
Fruit	1.96	0.80	2.13	1.89
Flowers	0.40	1.20	1.00	0.69
All categories	7.32	8.80	7.00	7.38

(1974) predicted for generalist herbivores. Presumably, both types of pressures are operating on howlers, resulting in more diversification of leaf species.

As pointed out earlier, the soft sugary pulp of many ripe fruits apparently functions to attract seed dispersal agents. Thus it is not surprising that howlers are able to feed heavily on one fruit species in a given day since toxins in most such foods should be relatively low. Also, since fruit sources are very patchy, howlers should be motivated to exploit such foods intensively wherever and whenever they find them.

As in the case of leaves, there was a positive correlation between the number of species eaten daily and time spent eating flowers ($r_s = .3064$, $p = .005$, $n = 85$). Flowers are even patchier than fruits with respect to time. But, unlike ripe pulpy fruits, flowers function to attract pollinators, not seed dispersal agents, and therefore may have evolved some chemical defenses against predators. Further, many flower sources were species of vines, which occupy much less space than the canopies of trees and therefore provide smaller patches of food than fruiting trees. Such patches would be fully exploited more quickly than patches of fruit, so more different sources would be required for the same amount of feeding time.

SPATIAL DIFFERENCES. There was no significant difference between the two areas in the number of species used daily (M-W, $p = .53$, $n = 40$, $m = 45$). Again, this suggests that despite the differences in the forest between the two areas, the pressures to diversify food species are similar.

TEMPORAL DIFFERENCES. In the combined data from both study areas, there were seasonal differences in the number of species eaten daily—most in the dry season, second in the wet, and least in the transition—but these differences were not statistically significant (K-W, $p = .0872$, $N = 85$); nor were such differences significant when the data were analyzed separately for each area. Thus at all times in an annual cycle the pressures to diversify food species are apparently similar, though phenological data show that at times there are many more species producing seasonal foods than at others. In spite of this, howlers adhere to their pattern of using 7 to 8 different food species per day.

Turnover of Food Species

OVERALL. Combined data from the two study areas show that there was a mean daily turnover of food species of 51 percent (see table 4.5). This is to say that about one half of the species eaten on one day were not eaten on the next. By this measurement, the mean daily turnover of species was 62 percent for leaves, 35 percent for fruits, and 71 percent for flowers (K-W, $p < .005$, $N = 49$). The daily turnover of food species can presumably be related to the duration of particular food categories.

Since flower crops have the shortest duration, it is not surprising that they have the highest daily turnover of species. The high turnover of young leaves may be attributed in part to the factors noted in the previous section—their generally lower estimated supply per individual tree and their generally higher content of toxins. But there is also evidence that in terms of optimal nutritional content, the duration of young leaves on particular trees is generally shorter

TABLE 4.5. Daily Turnover in Food Species

	Wet	Transition	Dry	Overall
Combined Data				
Leaves	.62	.32	.70	.62
Fruit	.36	.16	.39	.35
Flowers	.82	.55	.88	.71
All categories	.51	.31	.56	.51
Old Forest				
Leaves	.66	.37	.71	.65
Fruit	.43	0	.45	.40
Flowers	.75	.61	1.00	.72
All categories	.52	.33	.55	.51
Lutz Ravine				
Leaves	.59	.27	.68	.59
Fruit	.30	.31	.33	.31
Flowers	.88	.49	.75	.71
All categories	.51	.30	.57	.51

than that of fruits. (McKey 1975). Chemical analyses have shown that in some cases (e.g., *Ceiba pentandra*) the protein content of new leaf tips drops heavily, while the phenolic content and cell wall content rise rapidly within a period of a few days (Milton, unpub.). The soluble sugar content of fruits, however, increases with maturity (Spencer 1974). Ripe fruit may be available on particular trees in huge quantities for only a period of a few days (e.g., *Ficus* spp.). Or a given tree species may ripen only a portion of its crop or only a few fruit over a given period of time (e.g., *Spondias mombin* ripens many fruits simultaneously on a given tree over a period of weeks while *Virola surinamensis* ripens relatively few per day over a period of weeks).

There were no significant differences in daily turnover of food species between the two study areas or between seasons. It can be seen, however, that turnover of leaves was highest in the dry season, second in the wet, and least in the transition (K-W, $p < .10$, $N = 18$). This corresponds to the rank order of leaf production, as shown by the phenological data. Thus it appears that there is a greater turnover of leaf species when there are more species producing leaves.

TABLE 4.6. Monthly Turnover of Primary Food Sources

OLD FOREST		
Month	Species	Days Used in Sample
July	*Brosimum alicastrum*	4
	Poulsenia armata	1
Aug.	*Ficus yoponensis*	5
	Platypodium elegans	1
	Ficus costaricana	1
	Cordia alliodora	1
Sept.	*Poulsenia armata*	3
Oct.	*Quararibea asterolepis*	1
	Tabernaemontana arborea	1
Nov.	*Hyeronima laxiflora*	3
	Clusia odorata	1
	Ficus yoponensis	1
Dec.	*Inga fagifolia*	3
	Ceiba pentandra	2
Feb.	*Ficus insipida*	3
	Ficus yoponensis	3
	Arrabidea patellifera	1
	Cecropia insignis	1
	Pseudobombax septenatum	1
	Bignoniaceae sp.	1
March	*Ficus yoponensis*	2
	Ficus insipida	2
	Platypodium elegans	1
	Ficus costaricana	1
	Inga fagifolia	1
April	*Brosimum alicastrum*	4
	Ficus insipida	2
	Platypodium elegans	1
	Cecropia insignis	1

On a month-to-month basis, species turnover was 66 percent. This presumably reflects the different phenological patterns of different tree and vine species, which produce flush, fruit, and flowers at different times of the year. Primary food sources—food sources which accounted for 20 percent or more of feeding time in any given day—showed a very high monthly turnover rate, 78 percent (see table 4.6). This rate was 88 percent in the Old Forest and 67

TABLE 4.6. (*Continued*)

	LUTZ RAVINE	
Month	Species	Days Used in Sample
July	*Ficus insipida*	5
	Ficus yoponensis	2
	Platypodium elegans	1
	Poulsenia armata	1
	Maquira costaricana	1
Aug.	*Ficus insipida*	3
	Ficus yoponensis	3
	Luehea seemannii	1
Sept.	*Spondias mombin*	4
	Ficus yoponensis	3
	Inga fagifolia	2
Oct.	*Ficus yoponensis*	5
	Inga goldmanii	1
Nov.	*Ficus yoponensis*	5
	Ficus insipida	2
Jan.	*Ficus yoponensis*	3
	Ficus insipida	3
	Ceiba pentandra	1
	Inga sp.	1
Feb.	*Ficus insipida*	3
	Lacmellea panamensis	3
	Inga sp.	1
	Dioclea reflexa	1
	Ficus trigonata	2
March	*Ficus yoponensis*	5
	Pterocarpus rohrii	2
	Brosimum alicastrum	1
	Lacmellea panamensis	1
April	*Platypodium elegans*	5
	Ficus insipida	4
	Ficus yoponensis	1

percent in the Lutz Ravine. The lower turnover rate in the Lutz Ravine was due to the almost continuous use of fig products in this area.

Turnover of food species was also analyzed in terms of the number of consecutive days a particular species was eaten during a sample period. Most species were eaten on only one day in a sample

TABLE 4.7. Number of Consecutive Days in Sample Period that Particular Species Were Used as Food Sources

Combined Data	Consecutive Days					Relative Use				
	1	2	3	4	5	1	2	3	4	5
Leaves	149	33	11	5	4	.74	.16	.05	.02	.02
Fruit	32	12	7	7	6	.50	.19	.11	.11	.09
Flowers	27	4	1	0	2	.79	.12	.03	.00	.06
Wet Season										
Leaves	90	16	5	2	2	.78	.14	.04	.02	.02
Fruit	16	4	3	2	4	.55	.14	.10	.07	.14
Flowers	6	1	1	0	0	.75	.13	.13	.00	.00
Dry Season										
Leaves	51	11	2	3	0	.77	.17	.03	.05	.00
Fruit	15	7	3	5	1	.48	.23	.10	.16	.03
Flowers	18	2	0	0	2	.82	.00	.01	.00	.09
Transition Season										
Leaves	8	7	4	0	2	.38	.33	.19	.00	.10
Fruit	1	1	1	0	0	.33	.33	.33	0	0
Flowers	3	1	0	0	0	.75	.25	0	0	0
Old Forest										
Leaves	75	20	3	2	1	.74	.20	.03	.02	.01
Fruit	14	6	3	4	1	.50	.21	.11	.14	.04
Flowers	14	3	0	0	1	.78	.17	0	0	.06
Lutz Ravine										
Leaves	74	14	8	3	3	.73	.14	.08	.03	.03
Fruit	18	6	4	3	5	.50	.17	.11	.08	.14
Flowers	14	1	1	0	1	.82	.06	.06	0	.06

period, and relatively few were eaten on more than two continuous days (see table 4.7). In the food categories, species used as fruit sources were eaten two to five consecutive days far more often than species used as leaf or flower sources. These results show more of the pattern of species turnover than the simple rate of turnover, but they are related and may be interpreted in the same way.

The pattern was similar in both study areas, but there was a noticeable difference between the pattern of the transition season and that of the wet and dry seasons. During the transition season there was much more of a tendency to eat both leaf and fruit species for two to three consecutive days and even five consecutive days. At this time of year, when production of leaves and fruit is lowest, howlers may be forced to exploit the available food resources more intensively.

Differential Use of Species

OVERALL. The overall pattern in the differential use of food species, as measured in percent of feeding time, is that a few species were used rather heavily but most were used hardly at all (see table 4.8 and figure 4.2). The same pattern is seen if differential use of food species is measured by the number of days a species was eaten.

TABLE 4.8. Food Species Analysis by Study Area

	OLD FOREST			
Species	Categories Eaten	Percent of Feeding Time	Percent of Trees in Sample	Days Eaten
Ficus yoponensis	L, F	15.32	0.39	12
Brosimum alicastrum	L, F	11.25	0.59	9
Poulsenia armata	L	6.04	3.13	8
Ficus insipida	L, F	5.91	0.78	13
Inga fagifolia	L	5.18	0.40	8
Platypodium elegans	L, Fl	4.20	1.37	10
Cecropia insignis	L, F, Fl, P	3.44	0.98	13
Hyeronima laxiflora	L, F, Fl	3.23	0.98	6
Pseudobombax septenatum	Fl	2.21	1.17	7
Ficus costaricana	F	1.99	0.20	2
Quararibea asterolepis	L, F	1.86	5.28	4
Anacardium excelsum	L, F, P	1.79	0.39	10
Tabernaemontana arborea	L	1.64	0.39	4
Trichilia cipo	L, F	1.50	6.65	6
Eugenia oerstedeana	L, F	1.46	0.78	4
Ceiba pentandra	L	1.40	0.59	3
Arrabidaea patellifera	L	1.24	°	3
Bignoniaceae sp.	L	1.13	°	4
Clusia odorata	L, Fl	1.23	°	5
Chrysophyllum panamense	F	0.96	0.39	5
Dipteryx panamensis	F	0.91	0.98	5
Ormosia coccinea	L	0.78	1.57	4
Topobaea praecox	L, Fl	0.68	°	4
Abuta racemosa	L	0.67	°	2

TABLE 4.8. (*Continued*)

	OLD FOREST			
Species	Categories Eaten	Percent of Feeding Time	Percent of Trees in Sample	Days Eaten
Maripa panamensis	L, Fl	0.64	°	4
Acacia glomerosa	L	0.64	°	3
Cordia alliodora	L	0.55	1.96	1
Ficus obtusifolia	L	0.55	0	3
Tetragastris panamensis	L, F	0.50	1.76	2
Calophyllum longifolium	F	0.50	0	2
Protium panamense	L	0.48	0.78	1
Hiraea sp.	Fl	0.45	°	3
Paragonia pyramidata	L	0.42	°	3
Arrabidaea candicans	L	0.38	°	3
Inga goldmanii	L	0.37	0.98	3
Ficus trigonata	L, F	0.35	°	1
Mascagnia hippocrateoides	Fl	0.34	°	2
Ficus hartwegii	L, F	0.32	°	1
Souroubea sympetala	L	0.32	°	3
Drypetes standleyi	L, F	0.31	0	1
Ficus paraensis	L, F	0.31	°	3
Ficus citrifolia	L, F	0.28	0	1
Triplaris americana	L	0.27	0.39	1
Paullinia sp.	L	0.27	0	2
Tabebuia guayacan	Fl	0.27	0.59	1
Martinella obovata	L	0.27	°	1
Entada gigas	L	0.25	°	2
Ormosia sp.	L	0.25	0	1
Guatteria dumetorum	Fl	0.25	2.15	1
Tachigalia versicolor	L	0.25	0.39	1
Prioria copaifera	L, F	0.24	1.17	3
Ficus tonduzii	L, F	0.20	0.20	1
Machaerium arboreum	Fl	0.19	°	1
Cordia sp.	L	0.18	0	1
Desmopsis panamensis	L	0.15	0	1
Serjania sp.	L	0.15	°	1
Luehea seemannii	L	0.15	1.17	1
Inga sp.	L	0.15	0	1
Philodendron radiatum	P	0.13	°	2
Mangifera indica	F	0.13	0	1
Socratea durissima	F	0.11	0	1
Pterocarpus rohrii	L, F	0.10	0	1
Machaerium purpurascens	L	0.10	°	1
Licania platypus	L	0.09	0	2
Eugenia nesiotica	L	0.08	0.20	1
Celtis shippii	L	0.08	0.20	1

TABLE 4.8. (Continued)

	OLD FOREST			
Species	Categories Eaten	Percent of Feeding Time	Percent of Trees in Sample	Days Eaten
Alseis blackiana	L	0.06	5.09	2
Beilschmiedia pendula	F	0.03	1.37	2
Omphalea diandra	L	0.03	°	1
Tynnanthus croatianus	L	0.03	°	1
Virola surinamensis	L	0.03	2.74	1
Unonopsis pittieri	L	0.03	0	1
Hybanthus prunifolius	L	0.03	0	2
	LUTZ RAVINE			
Ficus yoponensis	L, F	25.95	2.15	35
Ficus insipida	L, F	22.88	1.16	33
Platypodium elegans	L, Fl	6.94	1.98	17
Spondias radlkoferi	L, F	4.96	1.65	10
Lacmellea panamensis	F	3.16	0.17	5
Pterocarpus rohrii	L, F, Fl	2.48	2.97	6
Inga fagifolia	L	2.04	0.17	6
Maquira costaricana	L, F	1.59	1.16	6
Brosimum alicastrum	L, F	1.48	0	6
Poulsenia armata	L	1.48	2.15	8
Ficus trigonata	L, F	1.41	0	3
Anacardium excelsum	L, F	1.17	3.80	10
Luehea seemannii	L	1.10	4.62	4
Inga sapindoides	L	1.07	0	2
Cecropia insignis	L, F, Fl, P	1.17	0.17	7
Inga sp.	L, Fl	0.93	0.1	3
Hyeronima laxiflora	Fl	0.90	4.62	4
Inga goldmanii	L	0.89	0.17	2
Inga punctata	L	0.87	0	3
Zanthoxylum panamense	L, Fl	0.82	0.83	5
Ceiba pentandra	L	0.81	0.17	2
Eugenia coloradensis	L, F	0.74	0	4
Protium panamense	L	0.77	0.99	4
Machaerium purpurascens	L	0.71	°	3
Dioclea reflexa	Fl	0.70	°	2
Socratea durissima	F	0.49	0	5
Entada gigas	L	0.44	°	2
Hasseltia floribunda	F, Fl	0.43	0.83	2
Trophis racemosa	L, F, Fl	0.43	1.98	3
Gnetum leyboldii	Fl	0.36	°	1
Cordia alliodora	L	0.36	0.66	3
Ficus tonduzii	L, F	0.32	0.17	4
Tetragastris panamensis	L, F	0.28	0	3

TABLE 4.8. (*Continued*)

		LUTZ RAVINE		
Species	Categories Eaten	Percent of Feeding Time	Percent of Trees in Sample	Days Eaten
Prioria copiafera	L, F	0.29	0	1
Quassia amara	L	0.24	0	1
Machaerium pachyphyllum	L, Fl	0.22	°	2
Arrabidaea patellifera	L	0.22	°	3
Bignoniaceae sp.	L, Fl	0.19	°	1
Cavanillesia platanifolia	L	0.19	0.17	1
Pseudobombax septenatum	Fl	0.19	0	1
Virola sebifera	L	0.18	2.81	3
Tabebuia guayacan	Fl	0.17	0	1
Miconia argentea	F	0.16	0	1
Maripa panamensis	L	0.13	°	2
Ormosia coccinea	L	0.12	0	3
Tetracera sp.	L	0.12	°	1
Terminalia amazonica	L	0.12	0.83	1
Inga sp.	L	0.11	0	4
Philodendron radiatum	P	0.11	°	2
Alseis blackiana	L	0.10	7.26	2
Hippocratea volubilis	L	0.09	°	3
Doliocarpus olivaceus	F	0.09	°	1
Calophyllum longifolium	F	0.09	0.17	1
Aspidosperma megalocarpon	L	0.07	0	1
Apeiba membranacea	L	0.06	0.99	1
Martinella obovata	L	0.05	°	2
Guatteria dumetorum	Fl	0.04	0	1
Phryganocydia corymbosa	L	0.04	°	1
Petrea volubilis	L	0.04	°	1
Clusia odorata	L	0.04	°	1
Bombacopsis sessilis	L	0.03	0.83	1
Tabernaemontana arborea	L	0.03	0	1
Macfadyena unquis-cati	Fl	0.02	°	1
Dipteryx panamensis	L, F	0.02	0.17	1
Gustavia superba	L	0.02	3.30	1
Triplaris americana	L	0.02	0.50	1
Connarus panamensis	L	0.02	0	1
Protium tenuifolium	L	0.02	0.99	1
Unonopsis pittieri	L	0.02	0.17	1
Ficus bullenei	L	0.02	0	1
Virola surinamensis	L, Fl	0.02	3.63	1
Piper arboreum	L	0.02	0	1
Doliocarpus major	F	0.02	°	1

Note: L = leaf; F = fruit, Fl = flower, P = petiole
° = vine, liana, epiphyte

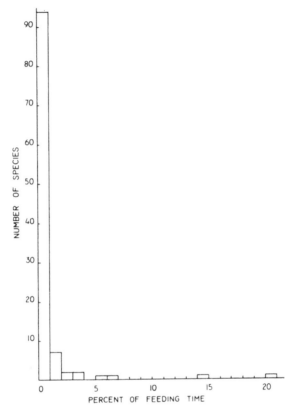

Figure 4.2.   Differential Use of Species
Number of species which accounted for >0 and ≤1% of overall feeding
time, etc., in combined data from both study areas.

A few species were eaten on many days but most were eaten on only one day during the entire sample period.

This pattern was noted in both study areas, but howlers in the Old Forest were somewhat more diversified than howlers in the Lutz Ravine, as measured by the Shannon-Wiener index (3.25 vs. 2.61). Since exactly the same number of species was used in both areas, the difference in the index is entirely due to the "evenness" component. The less even use of food species in the Lutz Ravine appears to be due primarily to the large percent of feeding time spent on *Ficus* products.

There were differences between the two study areas in the per-

cent of feeding time spent on the same species. The data were analyzed to see if these differences were related to differences between
the two areas in the relative density of the species. Of the 19 species
that were in the top thirty-three food sources for both study troops,
14 had higher percentages of feeding time in the area where they had
higher relative densities. This relationship was found to be statistically significant (phi coefficient = 0.55, $p < .01$, $n = 18$).

DAILY PATTERN.   A species that was used for 20 percent or more
of feeding time in a given sample day was considered a "primary"
food source while species used less than 20 percent of feeding time
were considered "secondary" food sources. Combined data from
both study areas show that howlers had a pattern of relying on one
or two primary food sources and five or six secondary food sources
each day (see table 4.9). There were only three days out of the total
of eighty-five sample days when no primary sources were used.
Overall, of the 132 cases of primary resource utilization, 50 percent
were fruit sources, 37 percent were leaves, and 8 percent were flowers (the rest were used for more than one food category). The general daily pattern suggests that whenever possible howlers selected
a primary food source, typically a fruit source, and fed heavily
from it one or more times in a given day. They supplemented this
primary resource with other food sources, generally leaf sources,
each of which was used less intensively.

Spatial differences: The same pattern of utilization of primary
and secondary sources was noted between areas. In the Old Forest

TABLE 4.9. Mean Daily Number of Primary and Secondary Food Sources

	Season			
	Wet	Transition	Dry	Overall
Combined Data				
Primary	1.44	1.30	1.77	1.54
Secondary	5.73	7.00	6.40	6.12
Old Forest				
Primary	1.30	1.00	1.67	1.40
Secondary	5.70	6.80	7.67	6.58
Lutz Ravine				
Primary	1.56	1.60	1.87	1.67
Secondary	5.76	7.20	5.13	5.71

there was somewhat less reliance on primary sources (1.40) than in the Lutz Ravine (1.67) and more reliance on secondary sources (6.58 vs. 5.71). Thus, on a daily basis, howlers in the Old Forest were also somewhat more diversified in their food sources than howlers in the Lutz Ravine.

Temporal differences: No significant differences could be detected in the number of primary and secondary sources used during the different seasons.

STAPLE SPECIES. A species that was used as a primary source in more than one sample month was considered to be a "staple" resource. As noted above, very few species could be regarded as staples because of the very heavy turnover rate of most howler food sources. Combined data from the two study areas show that only 9 species were used as staple foods, most frequently *Ficus yoponensis*, *F. insipida*, *Platypodium elegans*, and *Brosimum alicastrum* (see table 4.7). Of the 9 species, 2 were used only as leaf sources, 2 only as fruit sources, and 5 as sources of more than one food category.

In the Old Forest 8 species were used as staples compared with 4 in the Lutz Ravine. Again, in this respect, howlers in the Old Forest were more diversified in their food sources than howlers in the Lutz Ravine.

SPECIES USED ONLY ONCE. A considerable number of species were used as food sources on only one occasion during the study (42 percent). These sources tended to be leaf sources and flower sources. The ephemeral nature of many flowers may explain why they were eaten only once, but the onetime use of so many leaf sources cannot be explained simply in terms of ephemerality.

It is possible that these leaf species might be providing howlers with certain trace elements or rare amino acids. Such an explanation, however, appears improbable because of the great number of species utilized in this manner. It seems more probable that by this steady pattern of feeding only once, and for short periods of time, on many different leaf species, howlers may in fact be sampling potential leaf sources in their habitat and gauging their palatability or nutritional value as dietary items. It has recently been suggested that rats possess rapidly acting nutrient receptors located high in the upper gastrointestinal tract (Puerto et al. 1976). These specialized receptors apparently can very quickly recognize some of the nutritional components of foods being eaten and signal this information to the central nervous system. Howlers might possess similar feedback mechanisms allowing them to rapidly gauge some of the nutritional and/or toxic components of unfamiliar food items.

Such sampling behavior could serve an important adaptive function. As noted above, tropical forests are not static but dynamic communities that are constantly changing in species composition. Howlers that rigidly adhered to a circumscribed number of food species might, after several generations in the same area, find themselves in a nutritional cul-de-sac if their preferred food species were gradually being phased out. By frequently sampling many potential plant foods, howlers may be able to maintain dietary flexibility and take advantage of new sources of edible leaves, fruit, or flowers as they appear as members of the forest community.

Species Preferences

As explained earlier, if factors other than availability seemed to be influencing howlers in their choice of food, they were assumed to be showing a "preference." To determine if howlers were showing preferences in their choice of food species, I analyzed the time spent feeding on different species in relation to the spatial and temporal availability of food.

In the combined data from the two study areas, the top 10 food species, accounting for 63.59 percent of total feeding time, accounted for only 11.72 percent of the total trees in the combined sample plots (see table 4.10). In the Old Forest these figures were, respectively, 58.77 percent and 9.98 percent; in the Lutz Ravine, 72.96 percent and 13.56 percent. The disparities between feeding time and relative density indicate a preference for these species. But, to examine the question further, tests were done on the top 33 food species in each of the two study areas to see if there was any correlation between feeding time and relative density. There was no significant correlation in either area (O.F., $r_s = 0.15$, $p > .10$; L.R., $r_s = 0.09$, $p > .10$). This suggests that there is little relationship between the time spent feeding on a species and its relative density.

The top 10 howler food species were then compared with 10

TABLE 4.10. Food Preferences: Top Ten Food Species vs.
Relative Density in Combined Sample Plot Data

Species	Percent of Feeding Time	Relative Density (Percent)
Ficus yoponensis	20.95	1.34
Ficus insipida	14.89	0.98
Brosimum alicastrum	6.08	0.27
Platypodium elegans	5.65	1.70
Inga fagifolia	3.86	0.27
Poulsenia armata	3.63	2.60
Spondias mombin	2.63	0.98
Cecropia insignis	2.24	0.54
Hyeronima laxiflora	1.99	2.95
Lacmellea panamensis	1.67	0.09
Totals	63.59	11.72

species not eaten (or hardly eaten) by howlers with respect to the
number of months seasonal items had been shown to be available
on the trees by phenological data. Species were compared with
respect to leaves, fruit, and flowers by month; each category was
tested separately, and the categories were combined and tested
overall. The species howlers eat do not provide seasonal items in
significantly more months than the species they do not eat (M-W:
leaves, $p < .10$; fruit, $p = .10$; flowers, $p < .10$; overall, $p < .10$;
one-tailed tests, $n = m = 10$; see table 4.11).

The above evidence supports the view that howler food choices
are not determined primarily by the relative density of tree species
in the forest or by the number of months seasonal items are available
on the species. Other factors must therefore influence their food
choices.

ACCESSIBILITY AND/OR AVAILABILITY. One factor which may well
deter howlers from eating the products of certain tree species is
that, in general, howlers on Barro Colorado spend most of their
time in the middle to upper canopy. Thus they do not often come
into contact with understory trees. Of the 20 species listed in table
4.11, 5 of the 10 species not eaten by howlers are members of the
understory; none of the top 10 are understory species. Furthermore,
understory species tend to be small in comparison with upper canopy species and thus offer far less food per crop per tree. It may

TABLE 4.11. Species Preferences

	Number of Months Food Available			
	Young leaves	Fruit	Flowers	All seasonal foods
Top Ten Food Species				
Ficus yoponensis	12	12	—	12
Ficus insipida	12	12	—	12
Brosimum alicastrum	8	5	3	8
Platypodium elegans	5	4	1	6
Inga fagifolia	12	3	4	12
Poulsenia armata	5	2	4	7
Spondias mombin	9	5	3	12
Cecropia insignis	8	8	6	11
Hyeronima laxiflora	12	9	12	12
Lacmellea panamensis	9	3	1	11
Ten Species Not Eaten				
Hirtella triandra	10	4	4	10
Jacaranda copaia	7	5	2	7
Triplaris cumingiana	9	2	4	11
Palicourea guianensis	11	6	3	12
Ocotea curnua	6	3	3	6
Zuelania guidonia	4	3	3	7
Hura crepitans	5	3	3	7
Gustavia superba	7	5	5	8
Rheedia acuminata	9	0	2	10
Sterculia apetala	7	7	3	10

not be energetically feasible for howlers to seek out understory foods as a general rule when larger canopy food crops are also available. In addition, the relatively small branches of understory trees might not support the weight of these large monkeys.

NUTRITIONAL AND/OR NON-NUTRITIONAL FACTORS. A factor which should be of critical importance in determining whether a howler will eat a potential dietary item is the nutritional content of the item and the manner in which its particular nutrients (or other chemical constituents) combine with those of other foods available in the habitat.

In looking over the list of top 10 howler food species in table 4.11 and comparing their leaves, fruits, and flowers with those of the

Leaves of Ficus insipida *(Moraceae).*
A highly preferred leaf source for howlers on Barro Colorado Island. Animals will eat only the youngest rolled leaf in the center of the picture and possibly the upper portion of the next youngest leaf on this spray. More mature leaves will be ignored.
(Photo: K. Milton)

Leaves of Cecropia insignis *(Moraceae).*
Another highly preferred howler leaf source. Note the difference in texture between the young leaf on the left and the mature leaf on the right. Animals will eat the young leaf and ignore the older one.
(Photo: K. Milton)

Young leaf of Tachigalia paniculata *(Leguminosae).*
All of the leaflets on this one compound leaf are young. Note, however, that howler monkeys selectively eat only the tender tips of the leaflets, ignoring the more fibrous basal sections.
(Photo: K. Milton)

Fruits of Ficus yoponensis *(Moraceae), Eaten from Four Different Trees on the Same Sample Day.*
Howlers ate some fruit from each tree. The tiny young fruits contain more protein per unit dry weight while the older fruits contain more nonstructural carbohydrates (Milton, unpub.). By mixing fruits from several stages of maturity, howlers should obtain a better complement of essential nutrients than otherwise would be the case.
(Photo: K. Milton)

Adult male animal, giving the characteristic howling vocalization while completely suspended by the prehensile tail. Howling appears to function primarily as an intertroop spacing mechanism.
(Photo: Roy Fontaine)

Adult male howler contemplates a twig of very young leaves.
(Photo: George Angehr)

Subgroup of howlers in a characteristically relaxed pose. Howlers spend an average of 65 percent of their daylight hours resting, apparently as part of an overall strategy of energy conservation.
(Photo: Michael May)

species not eaten, certain broad trends indicative of nutritional biases of howlers can be discerned. Howlers ate young leaves of all but 2 species on the top 10 list (*Lacmellea panamensis* and *Spondias mombin*) but ate the mature leaves of only 2 (*Platypodium elegans* and *Ficus yoponensis*). Further, data show that howlers generally prefer to eat young leaves. This suggests that mature leaves as a class may contain constituents which make them unacceptable to howlers as food. The fact that howlers ignored both young and mature leaves from the 10 "not eaten" species may be partially because half are understory species. However, both young and mature leaves of the canopy species on this list were also ignored which suggests that certain young leaves too are unacceptable to howlers as food.

All species but one in the top 10 list produce fruits with either a soft pulp or an aril. The one species whose ripe fruits do not have a sugary pulp, *Platypodium elegans*, has wind-dispersed fruit. Of the 10 species not eaten by howlers, 2 (*Jacaranda copaia* and *Triplaris cumingiana*) have wind-dispersed fruit; 2 others (*Sterculia apetala* and *Hura crepitans*) produce fruit with a fairly hard pericarp and, in the case of *Sterculia apetala*, this pericarp is lined with stiff, irritating hairs, making it difficult to extract the seeds. *Gustavia supurba* produces a large fruit with hard seeds bound into a tough matrix that howlers, with their lack of manual dexterity, might find difficult to handle; *Ocotea curnua* produces fruit with a medium-sized seed and little edible pulp. *Rheedia acuminata* fruits every other year and was not fruiting at the time of this study; it is possible that howlers eat *Rheedia* fruit when it is available. Other than the fact that *Palicourea guianensis* and *Hirtella triandra* are members of the understory I can see no obvious reason why howlers might not eat the fruits of these species nor does there appear to be any obvious reason why howlers might not eat fruits of *Zuelania guidonia*, which is a canopy species. There is a strong probability that the omission of these trees from the list of howler food species represents undersampling. Overall, in the case of fruits, it appears that howlers prefer to eat those which are characterized by a fleshy mesocarp or rich aril—in short, species whose seeds are dispersed by animals and whose pulps and arils presumably function as attractants for dispersal agents.

With regard to flowers, few comments can be made. Many species included on the top 10 list are members of the Moraceae, a family characterized in general by extremely minute flowers. In the case of *Ficus* spp., the flowers are completely hidden inside syconia resembling small unripe fruits. Since to the eye, flowers that are eaten appear as available and edible as those ignored, chemical factors must be involved here.

NUTRITIONAL ANALYSES.   To determine more precisely what nutritional and/or other chemical factors might influence food preferences of howlers, analyses were carried out on a variety of plant parts from the Barro Colorado forest. Leaves, fruits, and flowers were examined for total protein, cell wall constituents, nonstructural carbohydrates, crude fat, and total phenolics including the presence of condensed tannin materials. Techniques used in these analyses are discussed in detail in Milton 1979a.

Since leaves are an important dietary class for howlers and potentially pose severe digestive problems, considerable attention was paid to their chemical constituents. Both young and mature leaves of some species were analyzed; the young leaves generally were eaten while the mature leaves were not. Some young leaves not eaten by howlers were also analyzed. In addition, tests were carried out on a variety of fruits and flowers; some species were eaten by howlers and others were not.

RESULTS OF LEAF ANALYSES.   As a class, young leaves contained significantly more protein per unit dry weight than mature leaves ($P < .025$, $N = 9$, Wilcoxon signed ranks test) and significantly less cell wall material ($P < .005$, $N = 9$, Wilcoxon test). Thus by concentrating primarily on young leaves, howlers are taking in more protein and less cell wall material per unit dry weight of leaf eaten (Milton 1979a). Since howlers have a limit to the amount of food they can process per unit time, it is important for them not to load the digestive tract with low nutrient/high fiber food. Data from feeding trials indicate that a mean of 20 h is required for a given meal to begin to pass out of the digestive tract of howler monkeys; thus each meal represents a considerable temporal commitment and young leaves would appear to provide a higher nutrient return in exchange for the time invested in digestion.

Two species whose mature leaves are eaten by howlers (*Platypodium elegans* and *Ficus yoponensis*) have features which set them apart from most mature leaves. Mature leaves of *P. elegans* are unusually high in protein while those of *F. yoponensis* have relatively high protein contents as well as a relatively low cell wall content. Thus these two species of mature leaf retain certain features generally characteristic of young leaves. Young leaves of species not eaten by howler monkeys, such as *Jacaranda copaia* and *Anacardium excelsum*, had protein contents higher than the mean protein content of young leaves howlers eat; however, these leaf species also had a cell wall content higher than the mean of young leaves eaten. This too supports the view that the protein/fiber ratio is an important factor in determining the preference of howlers for young leaves.

Both young and mature leaves were low in nonstructural carbohydrates (i.e., 3 to 4 percent dry weight), which suggests that leaves are not generally eaten for ready energy. Tests for total phenolics showed that some young leaves had equal or higher phenolic contents than their mature counterparts (Milton 1979a). Even so, young leaves were preferred as food and mature leaves were largely ignored. This suggests that in general the phenolic content may not be an important factor in determining the leaf choices of howlers, at least intraspecifically.

RESULTS OF FRUIT ANALYSES.   Only ripe fruits were analyzed; these were not examined for phenolics since it was assumed that most ripe fruits were low in toxic substances. All fruits analyzed were high in nonstructural carbohydrates when compared with leaves. Some fruits, however, such as those of *Ficus* spp., were much lower in ready energy components than certain other fruits (i.e., around 13 percent dry weight nonstructural carbohydrates for fig fruit as compared with estimates of as much as 50 percent or more for arils of fruits such as those of *Tetragastris panamensis*). In general, pulps and arils analyzed were low in protein and crude fat. Protein content was generally less than 5 percent dry weight while crude fat fell as low as 0.1 percent to 0.5 percent. Certain fruits, such as those of *Gustavia supurba*, were relatively high in crude protein (13 percent dry weight) while others, such as the nuts of *Scheelea zonensis* were quite high in crude fat (33 percent dry weight). Yet neither of these fruits was eaten by howler monkeys. As noted above, *Gustavia supurba* seems to require a high degree of manual dexterity to harvest and eat which might deter howlers from feeding on it; nuts of *Scheelea zonensis*, a Palmae, are contained in an extremely hard shell which howlers apparently cannot open. Thus potentially very rich fruits may be ignored as dietary items by howlers because of mechanical difficulties associated with extracting their edible parts.

RESULTS OF FLOWER ANALYSES.   Flowers were highly variable in nutritional content, more so as a group than either leaves or fruits. *Dioclea reflexa* flowers, for example, contained around 9 percent crude protein, 4 percent nonstructural carbohydrates, and 0.35 percent crude fat/dry weight while those of *Tabebuia rosea* contained 10 percent protein, 23 percent nonstructural carbohydrates, and 1.2 percent crude fat/dry weight. Yet the flowers of *Dioclea reflexa* were eaten by howlers and those of *Tabebuia rosea* were ignored. Flowers of *Tabebuia guayacan* with 25 percent crude protein, 14 percent nonstructural carbohydrates, and 3.7 percent crude fat were avidly eaten; *Jacaranda copaia* with flowers containing some 23 percent crude protein was ignored; while *Zanthoxylem panamense*, whose flowers contain around 20 percent crude protein, was eaten. Currently, I have no explanation as to why certain flowers species are eaten and others are ignored. In the few cases where flowers of different species have also been analyzed for cell wall components, this factor does not appear important in determining howler flower preferences since flowers they eat may be higher in these components than those they do not eat. In the case of flowers, it would appear that certain as yet undetected chemical factors, perhaps unusual amino acids, may at times be involved in determining which flower species howlers will use as food.

The above data show that there are considerable differences in the nutritional content of leaves, flowers, and fruits in the Barro Colorado forest. Some of these items are of decidedly higher quality than others because of higher proportions of certain important nutrients and/or lower proportions of structural materials and/or lower concentrations of secondary compounds.

Howlers seem generally to seek out higher quality foods and ignore those of lower quality, particularly mature leaves. Since howlers lack the elaborate fermentation chambers characteristic of many Old World primate folivores, they presumably have a lower threshold with respect to the amount of low nutrient/high fiber food they can process efficiently in a given time period. Most mature leaves would appear to be below this critical threshold (Milton 1979a).

Use of *Ficus*

OVERALL. Combined data from both study areas show that the genus *Ficus* (Moraceae) accounted for by far the largest percent of overall feeding time (38.68 percent). It was used as a food source for seventy of the eighty-five sample days (see table 4.12).

Two *Ficus* species, *F. yoponensis* and *F. insipida*, were by far the most important *Ficus* species, accounting for 35.93 percent of overall feeding time. One or both of these two species were eaten on sixty-five of the eighty-five sample days.

SPATIAL DIFFERENCES. In the Old Forest, fig products (flush, leaves, and fruit) accounted for 25.3 percent of overall feeding time and were eaten on twenty-six out of forty sample days. In the Lutz Ravine, the concentration on fig products was almost double; they accounted for 50.58 percent of overall feeding time and were eaten on forty-four out of forty-five sample days.

The difference between the two areas in daily feeding time spent

TABLE 4.12. Mean Daily Use of *Ficus* Products as Percent of Feeding Time

	Wet	Transition	Dry	Overall
Combined Data				
F. yoponensis	26.99	8.20	10.24	21.03
F. insipida	11.73	19.30	18.18	14.90
All Ficus	40.62	26.60	38.79	38.68
Old Forest				
F. yoponensis	21.83	0	12.29	15.52
F. insipida	0.42	1.80	14.59	5.91
All Ficus	25.86	6.20	30.93	25.30
Lutz Ravine				
F. yoponensis	31.11	16.40	20.48	25.93
F. insipida	20.78	36.80	21.76	22.89
All Ficus	52.42	53.20	46.65	50.58

TABLE 4.13. Use of *Ficus* by Food
Category (mean daily percent
of feeding time)

	Leaves	Fruit
Combined Data		
F. yoponensis	5.98	15.05
F. insipida	3.93	10.97
All Ficus	11.39	27.29
Old Forest		
F. yoponensis	2.24	13.28
F. insipida	3.45	2.46
All Ficus	7.06	18.24
Lutz Ravine		
F. yoponensis	9.31	16.62
F. insipida	4.36	18.53
All Ficus	15.24	35.34

on fig products is statistically significant (M–W, $p < .0001$, $n = 40$, $m = 45$).

Table 4.13 gives a breakdown of the use of fig fruit and leaves in each study area for the two most important *Ficus* species. Again, *F. yoponensis* and *F. insipida* accounted for almost all of the fig products eaten. It is interesting to note that although trees of *F. insipida* are four times as common in the Old Forest as trees of *F. yoponensis*, howlers concentrated much more heavily on the fruits of *F. yoponensis*. Data collected by Morrison (1975) on both *Ficus* species in the Lutz Ravine as well as my data from the Old Forest show that *F. yoponensis* produces more crops of fruit per annum than *F. insipida*. Nonetheless, the great disparity in numbers reduces the potential for an extra fruit crop in the Old Forest. In the Lutz Ravine there were 48 individuals of *F. insipida* and 71 of *F. yoponensis*. Here howlers do not appear to have a preference for *yoponensis* fruit over *insipida* fruit but they do show a preference for *yoponensis* leaves over *insipida* leaves—a preference also demonstrated for *yoponensis* leaves in the Old Forest.

An analysis of the nutritional and other components of leaves of these two *Ficus* species show that *F. yoponensis* produces a very unusual leaf. Both young and mature leaves of this species, unlike those of other tree species tested to date, including *F. insipida*, do not show notable differences in nutritional content depending on relative leaf maturity. Young leaves of *F. yoponensis* average around 37 percent cell wall material, 9 percent crude protein, and and 10 percent nonstructural carbohydrates/dry weight while mature leaves average 33 percent cell wall material, 12 percent crude protein, and 7 percent nonstructural carbohydrates. The slightly lower cell wall content in the mature leaf appears due to the fact that the young leaves are enclosed in stipules while the older leaves are not. Phenolic materials in both young and mature leaves of *F. yoponensis* are around 6 percent, and no evidence was found of condensed tannin material in leaves of either age. Leaves of *F. yoponensis* are also unusually high in crude fat, which averages around 7 percent in young leaves and 5 percent in mature leaves. This crude fat, however, may represent waxes on the leaf surfaces (fig leaves are generally notably glabrous) and thus not be nutritionally useful to howlers. By eating both young and mature leaves of *F. yoponensis* howlers are thus getting fair amounts of several important nutrients as well as a relatively low amount of cell wall material.

In contrast, young leaves of *Ficus insipida* contain 30 percent cell wall material, 11 percent crude protein, and 3 percent nonstructural carbohydrates as contrasted with 36 percent cell wall material, 9 percent crude protein, and 5 percent nonstructural carbohydrates in mature leaves. Both young and mature leaves of this species contain around 6 percent phenolic material, and both gave evidence of condensed tannins. Thus howlers may spend more feeding time on leaves of *F. yoponensis* than those of *F. insipida* because they can eat both young and mature leaves.

In chapter 3, it was pointed out that the Lutz Ravine has more than nine times the number of large free-standing *Ficus* trees than are generally found in other areas of the island. It was further noted that trees of the genus *Ficus* are intraspecifically asynchronous in their phenological patterns and that fruit crops may be produced at any time in an annual cycle while some new leaves are produced continuously. Many of the differences in diet between the two study areas appear directly related to the far greater number of fig trees in the Lutz Ravine and the far greater use of their seasonal products. Phenological data on the 124 *Ficus* trees in the 25 hectares making up the major portion of the Lutz Ravine show that fig products were available to howlers here in every sample month over four different annual cycles (Morrison 1975; Milton et al. in prep.). It is not an exaggeration to say that howlers here are living in a type of fruit orchard where fig fruits (and young fig leaves) are apparently always available from some tree in the area.

The continuous availability of these foods helps to explain why the Lutz Ravine howlers show a fairly even proportion of leaves and fruits daily as well as monthly. The greater variability in the proportions of these food categories in the Old Forest appears to be related to the fact that in this area there is no single tree genus which provides a continuous supply of seasonal foods. The peaks and valleys in food categories here are a much clearer reflection of the actual peaks and valleys in the production of seasonal items by canopy trees overall. In the Lutz Ravine the effect on the howler diet of the overall phenological pattern is "masked" by the unusually large number of fig trees and their regular use as food sources. The greater use of fig products by howlers in the Lutz Ravine also helps to explain why there was less monthly turnover of food species in this area.

The data were analyzed to see if there was a relationship between time spent eating fig fruit and the availability of such fruit, as indicated by phenological data. In the Lutz Ravine, on a monthly basis, there is no significant correlation between fig-eating and fig availability (r_s = .32, p < .10, one-tailed test, n = 9). In the Old Forest, however, there is a very strong correlation between fig-eating and fig availability (r_s = .74, p < .025, one-tailed test, n = 9). Thus fig-eating in the Old Forest appears to be much more dependent on supply than in the Lutz Ravine. This implies that there are limits on fig-eating in the Old Forest that are not apparent in the Lutz Ravine because of the large number of fig trees and the constant availability of fig fruit.

TEMPORAL DIFFERENCES.   In the combined data from both study areas there were no significant seasonal differences in feeding time spent on fig products (all together) or on fig leaves. There was, however, a significant seasonal difference in the use of fig fruit

(K-W, p = .049, N = 85). Time spent eating fig fruit was about equal in the wet and dry seasons, but was considerably lower in the transition season—the same pattern as was noted earlier for the category *fruit*. Phenological data show that in the Old Forest the transition season is a time of no fig fruit production. It is a time of relatively high fig fruit production in the Lutz Ravine. But during my five-day sample, fig fruit appeared scarce.

Within the two study areas there were some significant seasonal differences in the use of fig products. In the Old Forest this was due mainly to the lower percent of feeding time spent on both fruit and leaves of *Ficus insipida* during the wet season. In the Lutz Ravine, there was a great increase in time spent eating leaves of both *F. insipida* and *F. yoponensis* during the transition season—the time when overall leaf production is lowest on an island-wide basis. But *Ficus* trees, being asynchronous in their phenological cycles, do not follow the overall pattern, and some individuals were producing leaves at that time, providing howlers with a food that was otherwise relatively low in supply.

The important role of the fig in the diet of the Barro Colorado howler monkey is further illuminated by the relationship between the use of *Ficus* products and the number of species eaten on a sample day. There is a significant negative correlation between these two variables (r_s = −.246, p = .026, n = 85). This is to say, the greater the percent of time spent feeding on *Ficus* products, the fewer different food species are included in the diet.

Family Preferences

Combined data for the two study areas show that howlers used forty-one different plant families as food sources during the sample period (see table 4.14). In number of species, the families most widely used were the Leguminosae (16), the Moraceae (15) and the Bignoniaceae (8). Howlers ate no more than 1 or 2 species in thirty of the forty-one families. Various families which are represented on Barro Colorado by a number of different species, such as the Flacourtiaceae and the Rubiaceae, were hardly used at all as food species.

In percent of overall feeding time, the families used most by howl-

TABLE 4.14. Food Species by Family: Combined Data

Family	Species
Anacardiaceae	*Anacardium excelsum* *Mangifera indica* *Spondias mombin* *Spondias radlkoferi*
Annonaceae	*Desmopsis panamensis* *Guatteria dumetorum* *Unonopsis pittieri*
Apocynaceae	*Aspidosperma megalocarpon* *Lacmellea panamensis* *Tabernaemontana arborea*
Araceae	*Philodendron radiatum*
Bignoniaceae	*Arrabidea candicans* *Arrabidea patellifera* *Macfadyena unguis-cati* *Martinella obovata* *Paragonia pyramidata* *Phryganocydia corymbosa* *Tabebuia guayacan* *Tynnanthus croatianus*
Bombacaceae	*Bombacopsis sessilis* *Cavanillesia platanifolia* *Ceiba pentandra* *Pseudobombax septenatum* *Quararibea asterolepis*
Boraginaceae	*Cordia alliodora*
Burseraceae	*Protium panamense* *Protium tenuifolium* *Tetragastris panamensis*
Chrysobalanaceae	*Licania platypus*
Combretaceae	*Terminalia amazonia*
Connaraceae	*Connarus panamensis*
Convolvulaceae	*Maripa panamensis*
Dilleniaceae	*Doliocarpus major* *Doliocarpus olivaceus* *Tetracera sp.*
Euphorbiaceae	*Drypetes standleyi* *Hyeronima laxiflora* *Omphalea diandra*
Flacourtiaceae	*Hasseltia floribunda*
Gnetaceae	*Gnetum leyboldii*

TABLE 4.14. (*Continued*)

Family	Species
Guttiferae	*Calophyllum longifolium*
	Clusia odorata
Hippocrateacea	*Hippocratea volubilis*
Lauraceae	*Beilschmiedia pendula*
Lecythidaceae	*Gustavia superba*
Leguminosae	*Acacia glomerosa*
	Dioclea reflexa
	Dipteryx panamensis
	Entada gigas
	Inga fagifolia
	Inga goldmanii
	Inga punctata
	Inga sapindoides
	Machaerium arboreum
	Machaerium pachyphyllum
	Machaerium purpurascens
	Ormosia coccinea
	Platypodium elegans
	Prioria copaifera
	Pterocarpus rohrii
	Tachigalia versicolor
Malpighiaceae	*Hiraea sp.*
	Mascagnia hippocrateoides
Marcgraviaceae	*Souroubea sympetala*
	Luehea seemannii
Melastomataceae	*Miconia argentea*
	Topobaea praecox
Meliaceae	*Trichilia cipo*
Menispermaceae	*Abuta racemosa*
Moraceae	*Brosimum alicastrum*
	Cecropia insignis
	Ficus bullenei
	Ficus citrifolia
	Ficus costaricana
	Ficus hartwegii
	Ficus insipida
	Ficus obtusifolia
	Ficus paraensis
	Ficus tonduzii
	Ficus trigonata
	Ficus yoponensis
	Maquira costaricana

TABLE 4.14. (*Continued*)

Family	Species
	Poulsenia armata
	Trophis racemosa
Myristicaceae	*Virola sebifera*
	Virola surinamensis
Myrtaceae	*Eugenia coloradensis*
	Eugenia nesiotica
	Eugenia oerstedeana
Palmae	*Socratea durissima*
Piperaceae	*Piper arboreum*
Polygonaceae	*Triplaris americana*
Rubiaceae	*Alseis blackiana*
Rutaceae	*Zanthoxylum panamense*
Sapindaceae	*Paullinia sp.*
	Serjania sp.
Sapotaceae	*Chrysophyllum panamense*
Simaroubaceae	*Quassia amara*
Tiliaceae	*Apeiba membranacea*
	Leuhea seemannii
Ulmaceae	*Celtis schippii*
Verbenaceae	*Petrea volubilis*
Violaceae	*Hybanthus prunifolius*

ers were the Moraceae (51.54 percent) and the Leguminosae (15.80 percent). It can be seen that together these two families accounted for about two-thirds of overall feeding time and therefore played a major role in the howler diet.

Both families were used more heavily in the Lutz Ravine, but together they still accounted for 59.39 percent of feeding time in the Old Forest. Thus, despite the differences between the areas in the use of species, there was a similar pattern of concentrating on these two families for sources of food.

Several possible explanations, which are not mutually exclusive, are offered for this apparent specialization on the Moraceae and Leguminosae.

RELATIVE DENSITY.   Howlers may be concentrating on these two families because they have high relative densities in the Barro Colorado forest. It has already been shown, however, that there was not a close relationship between time spent feeding on different species and their relative densities. Showing this for families is more difficult since a number of the species in some families are lianas or vines rather than trees, and hence were not counted in the sample plots. In the case of the Moraceae, all but two of the species eaten by howlers were counted in the sample plots, and these two were epiphytic species of *Ficus*, which accounted for a negligible percent of feeding time. Thus the percent of time spent eating Moraceae (51.54 percent) can be compared with the relative density of species belonging to this family (8.77 percent). The disparity suggests that the very heavy use of this family by howlers cannot be satisfactorily explained by its relative density.

In the case of the Leguminosae, if the time spent eating lianas and vines is netted out, the feeding time for this family is 14.15 percent versus a relative density of 8.56 percent. There is still a disparity, but here there is a much closer relationship between feeding time and relative density.

PHENOLOGY. Howlers may be concentrating on these two families because they produce seasonal foods more often or over a longer period of time than other families. Again, it has already been shown that the 10 species most heavily used as food sources by howlers do not produce seasonal foods in significantly more months of an annual cycle than species not used (or hardly used) as food sources. Such an analysis could not be done for families because of a lack of phenological data. But when the species used in the earlier test were examined, it was found that 7 of the top 10 food species belonged to the Moraceae or Leguminosae and 7 of the other group of species belonged to families that howlers did not use as food sources. When these two groups of 7 species each were compared, there was no significant difference in the number of months in which leaves, fruits, flowers, or seasonal foods all together were produced during an annual cycle.

The Moraceae, however, do have a phenological feature, at least in some species, that could help to explain why howlers use them so heavily. Many Moraceae, particularly *Ficus*, are intraspecifically asynchronous in their phenological cycles and further produce more than one crop of leaves and fruit per year. Thus by specializing to some extent on Moraceae products, howlers may be less susceptible to the overall fluctuations in the supply of preferred seasonal foods, especially on the downside.

Moraceae may thus provide a type of "floating" reserve of seasonal foods that help bring howlers through times when species in many other families are not producing seasonal foods in certain categories. *Cecropia insignis* and *Ficus* spp. produce some new leaves continuously and at times have large new leaf crops while *Poulsenia armata* produces flush several times per annum (Milton unpub. data). Three of these species also typically produce more than one fruit crop per annum and in the case of *Ficus* spp. such crops are intraspecifically asynchronous. Many members of the Moraceae produce similar animal-dispersed fruits and they often have somewhat different fruiting cycles, possibly to avoid competition for dispersal agents. *Brosimum alicastrum* puts out its biggest fruit crop in the early rainy season and produces another smaller crop toward the end of the rainy season. *Ficus insipida* and *F. yoponensis* are likely to produce fruit at any month during the year, but both species show strong peaks of fruit production in the dry season, although in different months of the dry season.

As it is such a large family, members of the Leguminosae also show a variety of different phenological strategies. Howlers use the Leguminosae as sources of flush leaves and flowers rather than as fruit sources. Further, many members of the Leguminosae eaten by howlers are lianas rather than trees. This is also true of the Bignoniaceae. By using legume and bignone products, howlers are "hedging" in another sense as the sheer number of species in these families, with their wide variety of phenological strategies, amplified by the fact that both tree and liana lifeforms are involved, make it possible that at any given time one or more of them may be producing young leaves and/or flowers.

NUTRITIONAL CONTENT.   Howlers may be concentrating on the Moraceae and Leguminosae because the foods provided by them have a relatively high nutritional content. As yet there are not sufficient data to support or refute this hypothesis. Analyses show that in some cases certain species produce relatively quite rich dietary items (e.g., young leaves of various members of the Leguminosae, mature leaves of *Platypodium elegans*, contain considerable protein). In other cases, however, foods from these families contain lower amounts of certain important nutrients than foods from other families (e.g., fruits of *Ficus* spp. are relatively low in nonstructural carbohydrates when compared with those from various other families on Barro Colorado). It would appear that a combination of nutritional and other factors, such as relative availability and/or toxicity may be involved in determining preferences for foods from particular families.

SECONDARY COMPOUNDS. Though plant species and even plant parts from the same species can apparently be quite variable in the types and amounts of secondary compounds they contain, there are sets of compounds that may be shared in common by most or all members of particular families (Levin 1971, 1976; Bell 1972; Fowden 1974). The study of these family-specific, genus-specific and species-specific chemical compounds has led to the development of plant chemosystematics which attempts to arrange plants phylogenetically by means of such traits (e.g. Hegnauer 1963).

Yet, as has been pointed out, what is highly toxic to one animal species may not be toxic to another (Freeland and Janzen 1974; and others). Apparently during their evolutionary history and/or lifetime, particular animal species have evolved mechanisms (biochemical and/or morphological) that help them to cope with particular sets of secondary compounds that many other species avoid (e.g., koala and eucalyptus leaves). It would probably not be advantageous to evolve detoxication mechanisms for foods from plant families that are not widely represented in space and time. It might, however, be highly advantageous to be able to deal with sets of compounds common to families that are widely represented. This is not to imply that the full range of secondary compounds shared by even a small plant family could necessarily be degraded by one animal species—but animals might be able to evolve mechanisms which would cope efficiently with some sets of compounds in particular families, at least up to a certain concentration.

Howlers may have evolved this strategy of specialization on certain plant families. The families howlers prefer are represented by a large number of species and are found throughout the Neotropics. By focusing their detoxication strategy on the common properties of some members of these families, howlers may have evolved an efficient method of dealing with some aspects of the toxic problem.

Species of the Moraceae provide howlers with both leaves and fruit. Many members of the Moraceae are characterized by the presence of latex, which is a terpene and hence might contain certain components not desirable to eat. It would seem that howlers are able to detoxicate terpenes or other secondary compounds characteristic of the Moraceae efficiently enough so that various latex-containing Moraceae products such as *Ficus* spp. leaves and unripe fruits can be heavily exploited. Ripe fig fruit should be very low in toxic materials as the small seeds of *Ficus* are dispersed through the intestines of the many animals eating the fruit. An understanding of the interrelationship between howler monkeys and secondary compounds characteristic of their foods is hampered at present by a lack of data both on Neotropical plant toxins and on the detoxication abilities of howler monkeys. There is some evidence to suggest that New World monkeys as a group may be considerably less efficient at degrading certain secondary compounds than Old World monkeys (Williams 1971). It is very difficult to get such data on particular primate species; further, it is extremely difficult to isolate many secondary compounds from plant parts. It is therefore likely that, for the time being, the relationship between particular primate species and the chemical components of particular plant families will remain largely speculative.

SUMMARY

1. During my study, howlers on Barro Colorado spent an overwhelming percent of feeding time on seasonal foods—young leaves, fruits, and flowers. There was evidence that they preferred these foods to perennial foods, such as mature leaves. As noted in chapter 1, seasonal foods are generally of a higher nutritional quality than perennial foods.

2. Typically, howlers ate substantial proportions of both leaves

and fruit each day. Leaves and fruit were evidently used as comple-
ments in the diet, leaves being the principal source of protein and
fruits being the principal source of ready energy. Flowers apparently were used both as sources of ready energy and of protein.
There was evidence that howlers preferred to eat both leaves and
fruit (and/or flowers) each day, presumably to get the required balance of energy and other nutrients.

3. Howlers ate foods from a great number of different species—
a minimum of 109 in the overall sample and 7 to 8 on a daily basis.
Most of these different species, both overall and daily, were used
as leaf sources. Since young leaves have lower contents of ready energy and are usually less concentrated than fruit (i.e., the potential
rewards for finding them are less), the pressure to diversify leaf
sources (and thereby reduce the costs of foraging) should have been
stronger. The pressure to diversify leaf sources to avoid an overload
of toxic compounds should also have been stronger.

4. There was a high daily turnover in the species used as food
sources. Leaf species had a much higher turnover rate than fruit
species, and flower species even higher. These differences correlate with the supply and duration of the three food categories on
individual plants. The monthly turnover in food species was even
higher than the daily rate. This reflects the different phenological
strategies of the plant species used by howlers as food sources.
Throughout the study period, howlers changed the species used as
food sources and continued to eat mainly seasonal items.

5. A few food species were used rather heavily (i.e., more than
10 percent of overall feeding time), but most were used hardly at
all. In fact, most were used on only one day of the sample period.
In terms of daily feeding time, fruit species were exploited more
intensively than leaf (or flower) species. There was a daily pattern
of having 1 or 2 primary food sources (20 percent or more of feeding
time) and 5 to 6 secondary sources. The primary sources changed
from month to month, and only a few could be considered "staples"
(i.e., were used as primary sources in more than one month of the
study). Many species were eaten in only minute amounts, and possibly they were being sampled as potential food sources.

6. An analysis of the chemical constituents of certain leaves, fruits,

and flowers showed that the seasonal foods howlers preferred were generally of a relatively high nutritional quality. The protein/fiber ratio seemed to be the most important factor in determining howler leaf choices. Fruits eaten were generally characterized by a soft pulp or aril and a high nonstructural carbohydrate content when compared with that of leaves. Flowers varied in their nutritional content, depending on the species analyzed. Phenolic content per se did not seem important in determining most leaf or fruit preferences but may have been important in some cases of flower selection.

7. A very strong preference was shown for two species of the genus *Ficus*, which together accounted for 35.93 percent of overall feeding time. Up to a point, more food was eaten from these two species where they were more abundant; indeed, in one of the two study areas there was a high degree of specialization on these two species. As both species are intraspecifically asynchronous in their phenological cycles, in the Lutz Ravine, where they are unusually abundant, they provide a virtually continuous supply of seasonal items, both young leaves and fruit. Here the pressures to diversify and change food sources are less strong than in the Old Forest, and this is reflected in a diet that is less diverse and less variable in terms of food categories and food species.

8. There was evidence that howlers preferred to eat items from certain plant families. In some cases, such items may have higher proportions of certain nutrients than potential foods that howlers ignore. They may also have classes of secondary compounds that howlers are able to tolerate or detoxicate more easily than those from certain other plant families. The most preferred families, the Moraceae and Leguminosae, are represented by a large number of species on Barro Colorado and are ubiquitous throughout the Neotropics.

5 RANGING

SINCE HOWLERS have been demonstrated to prefer seasonal foods, which are patchily distributed in both space and time, they have the continual problem of how to locate such resources. Unlike many secondary consumers, howler monkeys do not have to pursue their food once it is located; on the other hand, they cannot use the "sit and wait" strategy on the chance that edible food will suddenly appear. Howlers must travel to locate their food which, of course, involves an expenditure of energy.

In models of foraging strategy, an efficient strategy is expected to maximize the net return of energy from time or energy spent in foraging (Schoener 1971). Since a major food category in the howler diet (leaves) is extremely low in ready energy, the potential energy return from foraging is often limited. The howler digestive tract can hold only a limited amount of food at any one time and can be filled only a limited number of times per day. Therefore, to maximize their net return from foraging, howlers might be expected to minimize the energy spent in travel to locate food. They might be similar to what Schoener (1969, 1971) has called "time minimizers"— i.e., animals whose fitness is maximized when time spent foraging to gather a given energy requirement is minimized. In this context, however, the term "travel minimizer" seems more appropriate.

Presumably, the less an animal travels, the less food energy it requires—up to a point. Therefore, by minimizing travel, an animal might minimize its energy needs. The result might be a higher net energy return from a given supplying area.

Travel could be minimized by a search strategy that maximized the probability of finding preferred foods in relation to the distance traveled. Different search strategies would be optimal depending on the distribution of the preferred foods.

If the distribution of foods were uniform in space and time, the most efficient strategy would be to cover the entire supplying area in what might be called the "lawn mower" pattern, i.e., moving back and forth in uniform swaths until the total area was covered.

If the distribution of foods were random in space and time, the best strategy would be to travel about at random, since the probability of finding food would be equally good in all places at all times.

If the distribution of foods were patchy in space and time, the best strategy would be goal-directed travel, i.e., to move as directly as possible to sources of preferred foods where and when they were available.

Since I had demonstrated that the preferred foods of howlers are very patchy in space and time, I expected that they would use the third strategy. At the same time, since howler foods vary in the degree of patchiness (e.g., leaves vs. fruit) and since some species are very rare or were only used on one occasion, I also expected a certain amount of random movement within a basic strategy of goal-directed travel.

Therefore, in collecting data on howler ranging behavior, I wanted to know, on a daily basis, how the monkeys found sources of preferred foods and how much it cost them in terms of travel; and, over an annual cycle, how big a supplying area they required, how much of this area was utilized, and how exclusive it was with respect to other howler troops. To answer these questions I needed specific data on travel patterns, and time, distance, and rate of travel. Such data could then be analyzed in relation to data on food sources and feeding behavior.

METHODOLOGY

On beginning a five-day sample, I would mark the area where the howlers were sleeping with a dated piece of flagging. Then as the animals began to travel, I would follow their route, marking trees they passed through with other numbered, dated pieces of flagging. At the end of the five-day sample, I returned to the area with an assistant and, using the flagging as a guide, measured the routes with a metered surveyor's tape and Suunto compass. All data were recorded with reference to a baseline point in each of the two study areas and were further referenced by being superimposed onto prepared maps of the principal trail networks in each of the two study areas. Ultimately, the data were computerized and the ranging patterns (daily, monthly, and overall for each area) were drawn by a Hewlett-Packard 9830 model.

As the animals traveled, I noted their movements as part of the five-minute sampling program. For this purpose, any time an animal was in the process of moving from one tree to another its behavior was recorded as *travel*. Thus at the end of the day I had a record of how much time had been spent traveling.

I had daily distances traveled (from the measurements taken on the travel routes), and I calculated travel rates by dividing distances by times. All of these data—on time, distance, and rate of travel—were analyzed on a daily basis in the same manner as the data on feeding behavior. Spatial and temporal differences were tested for significance using Mann-Whitney or Kruskal-Wallis tests, as appropriate. Relationships between these data and other sets of data (e.g., on feeding behavior and food availability) were detected and tested using Spearman rank correlation tests. The results of these tests were considered to be significant at levels of .05 or lower. Unless otherwise indicated, the p values given for Mann-Whitney and Spearman rank tests are for two-tailed tests. Most of the tests were done on a CDC 6600 computer.

RESULTS

Travel Bouts

My study troops did not have a particular sleeping grove or sets of sleeping trees to which they returned each evening. They settled down for the night in whatever trees were most convenient in the area where they had been feeding prior to nightfall. Nor did they travel to get water (which they obtained from their food) or to find mates (their mating partners were in the troop). Thus it appeared that their main reason for traveling was to locate and/or monitor food sources.

Travel was generally instigated by the alpha male of the troop. This male would give a low, almost inaudible, cough, and the other members of the troop would be alerted and would gather in one area of the tree. Then in single file they would begin to leave the tree and move off, apparently in the direction indicated by the alpha male.

Generally the monkeys would travel through the trees in single file, a fact noted by earlier researchers (e.g., Carpenter 1934). Thus each monkey would take approximately the same route, both horizontally and vertically, as the other members of its troop. This suggests that travel was goal-directed; if the monkeys had been simply traveling through the forest at random, searching for food, presumably it would have been far more efficient for them to travel spread out in a phalanx such as has been reported for savanna-living baboons (Altmann and Altmann 1970) and arboreal grey-cheeked mangabeys (Wasser and Floody 1974). In that way a far greater area would have been covered. If travel were goal-directed, however, it would be more efficient to move in single file.

There is also a safety factor to consider. Many tropical trees have brittle or dead limbs, unreliable as supports for a relatively heavy animal such as the howler. Personal observations indicate that howler monkeys (and spider monkeys) generally fall not because they miss or lose their grip, but because the limbs they are using for support break off the tree. These limbs are not necessarily dead, but may simply be too brittle to support the weight of such large monkeys. (On two occasions, I narrowly escaped being hit by a falling branch, which broke under the weight of one to four feeding howlers. In both cases the branches were healthy and covered with flush leaves.) By traveling in single file, all animals use the same, proven route. If each animal moved along a different route, the probability of encountering rotten or brittle limbs would be proportionately higher.

Travel would generally lead to a food source, usually a single tree, where the members of the troop would deploy themselves and feed. The combined data show that 90 percent of all travel bouts were directly followed by feeding bouts (where a "bout" is defined as a half-hour period and all consecutive such periods in which 25 percent or more of the time was spent at the particular activity). This evidence supports the view that howlers traveled primarily to get to food sources.

Travel Patterns

PIVOTAL TREES.   A study of the travel patterns for individual five-day samples shows that they were generally oriented around one or two "pivotal" trees, which were always primary food sources (see table 5.1). More often than not, pivotal trees were fruit sources, but occasionally they were leaf or flower sources. Not unexpectedly, the species most often used as pivotal trees were *Ficus yoponensis* and *F. insipida.*

The data show that on a given day the animals would feed in such a tree, then leave it and feed elsewhere. They would return to the tree later in that same day and/or later in the same sample period. Howlers generally visited such trees repeatedly during the sample period, either leaving and returning by the same route or looping around and returning by a different route. The pivotal tree seemed to give the monkeys a base from which they could move out in various directions and search for other sources of food. As the pivotal trees were usually a fruit source, other sources of food were most likely to be flush leaves. A fruit source would provide monkeys with ready energy to travel and locate leaf sources, which are less patchy in space and time than fruit sources and hence would be more easily encountered; and while foraging in the area, the monkeys would have a *known* fruit source available, to which they could return as needed. By the time the food in the pivotal tree was depleted past the point of optimal harvesting or matured past the point of optimal edibility, the monkeys would have located new pivotal trees so that the same pattern could be repeated.

Travel patterns for the five-day samples show that to move from one pivotal tree to another, or to return to a pivotal tree from a secondary source, howlers often used the same arboreal route. This also supports the view that howler travel is goal-directed. If howlers were simply moving about at random, they would not repeatedly use the same routes. It could be argued that howlers use these routes because they are the only ones available—i.e., that the crowns of

TABLE 5.1. Pivotal Trees

OLD FOREST

Month	Category	Species	Tree	No. of Visits	No. of Days
July	fruit	*B. alicastrum*	Drayton	8	4
Aug.	fruit	*F. yoponensis*	62-I	10	5
Sep. Oct.[a]	leaf	*P. armata*	Wheeler	6	3
Nov.	flower	*H. laxiflora*	A–Z area	3	3
Dec.	leaf	*I. fagifolia*	#935	6	4
	leaf	*C. pentandra*	#987	6	4
Feb.	flower	*P. septenatum*	1, 2, 3	10	5
	fruit	*F. yoponensis*	62-I	3	3
March	fruit	*F. insipida*	#36	5	3
	leaf	*F. yoponensis*	Pair	3	2
April	fruit	*B. alicastrum*	S.C.	6	4
	leaf	*F. insipida*	E.S.	7	4

LUTZ RAVINE

Month	Category	Species	Tree	No. of Visits	No. of Days
July	fruit	*F. insipida*	T-37	8	5
	fruit	*F. insipida*	T-36	5	2
	fruit	*F. yoponensis*	T-9	2	2
Aug.	fruit	*F. insipida*	T-36	5	3
Sep.	leaf	*F. yoponensis*	T-13	8	5
	fruit	*S. mombin*	T-196	7	5
	leaf	*I. fagifolia*	Grove	5	5
Oct.	fruit	*F. yoponensis*	T-46	5	3
	leaf	*F. insipida*	T-41	3	3
Nov.	fruit	*F. yoponensis*	T-38	4	2
	fruit	*F. insipida*	T-41	4	2
Jan.	leaf	*F. yoponensis*	5-6 trees	9	4
	leaf & fruit	*F. insipida*	5-6 trees	8	5
Feb.	fruit	*L. panamensis*	Dock	5	3
	leaf	*F. trigonata*	S-103	3	3
March	fruit	*F. yoponensis*	T-46	12	4
	fruit	*F. yoponensis*	T-104	6	3
	fruit	*F. yoponensis*	T-105	9	5
	leaf	*P. rohrii*	T-219	6	4
April	fruit	*F. insipida*	Fig. X	7	4
	flower	*P. elegans*	#25, T-274 + 2 others	7	5

[a]No pivotal trees—only three days of data

particular trees touch only in certain areas, making travel by other routes impossible. This may be true in a few cases, but in most cases there appeared to be a number of different routes that could have been taken. Usually the route taken appeared to be the most direct, i.e., in as straight a line as possible.

ARBOREAL PATHWAYS. After studying each troop for several months, I became familiar enough with many of their travel routes so that I could predict which routes would be used to get from one area to another. Some of the more important of these routes have been named and their use quantified (see table 5.2). I regarded a particular travel route as an "arboreal pathway" if it (a) went a distance of approximately 100 meters or more, (b) connected important pivotal tree areas, and (c) was used more than once. For example, the Dock Hill to T-46 arboreal pathway in the Lutz Ravine area was used repeatedly by howlers, both during the sample months and at other times. Dock Hill is an important supplying area for howlers as it contains a *Lacmellea panamensis* tree that produces fruit over a period of months in the dry season and further is in close proximity to several large *Ficus* spp. trees and a huge *Anacardium excelsum*. The T-46 area is important because it contains various large *Ficus* spp. trees (including T-46 and T-47) and is also near several *Inga* spp. trees, a *Virola surinamensis*, and a much-utilized *Pterocarpus rohrii*.

TABLE 5.2. Some Major Arboreal Pathways

OLD FOREST	
Pathway	Times Used
Pseudobombax #815 to *Ficus* #62-1	10
Shortcut Loop	6
Pseudobombax to Swamp Loop	4
LUTZ RAVINE	
Pathway	Times Used
T-36 to T-196 area	14
T-46 to T-105	10
Dock Hill area to T-46	8
D-3 to T-46 area	6

Since these arboreal pathways connect particular trees or groves of trees that were used repeatedly as food sources during the sample period, it is likely that they were being used by the monkeys to locate food. Of course a pathway would lead only to a place where food might be available, but not necessarily at the right time; and in fact at times the monkeys passed through pivotal trees without feeding. But since many of these trees were fig trees, from which the monkeys ate both leaves and fruit, and since individual fig trees often produce new leaves, there would usually be at least some food available; and by regularly visiting such trees, the monkeys would be able to monitor them as potential food sources. As evidence that they were visiting such trees deliberately, it can be noted that in the Lutz Ravine on every sample day, the troop passed through at least one *Ficus yoponensis* or *F. insipida* tree, on most days through several, and on some days through 10 or more. Since these *Ficus* species account for only an estimated 3.31 percent of all trees in the Lutz Ravine (based on data from the three sample plots = 30,000 m²), it is not very probable that the monkeys would have passed through such trees so regularly and so often if they had simply been traveling at random.

Time, Distance, and Rate of Travel

Combined data show the howlers traveled a mean number of 1.23 hours per day and a mean distance of 443 meters. The mean rate of travel (distance/time) was thus 360 meters per hour (see table 5.3).

The minimum distance traveled in any given day was 104 meters and the maximum was 792 meters. As the standard deviation was only ±148, one can say that howlers generally travel from 300 to 600 meters per day. Thus, they do not usually remain in one place for several days and then travel a great distance (1,000 meters or more) to a new section of their home range, as suggested by Carpenter (1934), Richard (1970), and Schlichte (1978). Rather, they seem to have a pattern of regular daily travel, though at times such travel may be tightly oriented around one pivotal tree for two or more days. This pattern of regular daily travel might seem to contradict my hypothesis that howlers would be travel minimizers. By traveling a certain amount each day, they might not be minimizing

TABLE 5.3. Time, Distance, and Rate of Travel

Combined	Wet	Transition	Dry	Overall
Time (hrs)	1.17	1.31	1.29	1.23 ± 0.44
Distance (meters)	410	376	515	443 ± 148
Rate (meters per hour)	350	287	399	360
Old Forest				
Time (hrs)	1.10	1.25	1.38	1.22 ± 0.46
Distance (meters)	343	326	480	392 ± 127
Rate (meters per hour)	312	261	348	321
Lutz Ravine				
Time (hrs)	1.23	1.37	1.19	1.23 ± 0.42
Distance (meters)	464	426	550	488 ± 152
Rate (meters per hour)	377	311	462	397

travel for any given day; but, by traveling enough each day to monitor and exploit food sources within a certain area, howlers may well be minimizing travel over a longer period of time, since the major part of their travel might then be goal-directed rather than exploratory. By expending small amounts of energy regularly in short exploratory trips, especially at times when they have abundant sources of energy-rich foods, howlers may actually be saving energy over the long run—i.e., by avoiding *long* exploratory trips.

This is not to imply that howlers leave rich food sources with the intention of returning to them later. The pressure to leave such sources must come from the need for other nutrients to balance the diet. So howlers are evidently forced by their dietary requirements and the content and distribution of their foods to move a certain distance each day. But while they are moving about in an area, they do often return to known food sources, as already shown.

SPATIAL DIFFERENCES. There was no significant difference between the two areas in time spent traveling (M–W, $p = .266$, $n = 40$, $m = 45$), but there were significant differences in the distance traveled ($p = .0015$) and rate of travel ($p = .0314$). Howlers in the Lutz Ravine traveled longer daily distances and faster than howlers in the Old Forest. This may relate to the greater number of fig trees in the Lutz Ravine and the way such trees are distributed there. My analysis of fig tree distribution in thirty-two hectares of this area shows that here both *F. yoponensis* and *F. insipida* are significantly clumped on the scale of one hectare (variance/mean test, see table 5.4). (However, when data from the Lutz Ravine sample plots and Old Forest sample plots were combined, *F. insipida* did not show a significant tendency to clump on any scale. See table 3.5). Eleven of the thirty-two 1-hectare quadrats had no individuals of either species, while a few had 8 to 12 individuals. Though some of the latter quadrats were adjacent to each other, they were spread out across the supplying area and therefore, to reach a particular patch of fig trees, the monkeys often had to travel as much as 400 meters. Since such trips would be rewarded by unusually rich patches of food, there would be more incentive to travel in the Lutz Ravine than in the Old Forest, where there is only one such patch of fig trees. The greater travel rate of howlers in the Lutz Ravine also suggests that their travel is more goal-directed, as they move faster through the trees than howlers in the Old Forest. On many occasions, my study troop would set out from one side of the Lutz Ravine and cross to the other side, traveling directly to patches of fig trees more than 300 meters away, usually without stopping to feed along the way.

TEMPORAL DIFFERENCES.   Combined as well as separate data from the two study areas do not show significant seasonal differences in daily time spent traveling (K-W, $p = .47$, $N = 85$), but do show significant differences in distances traveled and travel rate (K-W, $p = .0004$; $p = .0001$). Distance and travel rate are greatest in the dry season, second in the wet, and least in the transition. This corresponds to the rank order of overall availability of seasonal items. Thus it would appear that howlers are traveling further and faster when the potential rewards are greater—which would be consistent with my expectations. At such times, the pressure to minimize travel would presumably be less strong.

On a monthly basis, there were positive correlations between the mean daily distance traveled and the availability of young leaves, fruit, flowers, and seasonal items all together, but in the combined data and in the Lutz Ravine, separately, none of these correlations were statistically significant. In the Old Forest, however, three of them were significant: distance traveled and availability of young leaves ($r_s = .75$, $p < .025$, one-tailed test, $n = 7$), fruit ($r_s = .68$,

TABLE 5.4. Distribution of *Ficus yoponensis* and *Ficus insipida* in the Lutz Ravine

Hectares	F. yoponensis	F. insipida	Total
S-1, W-2	8	4	12
N-0, W-1	8	2	10
N-0, E-0	3	5	8
S-3, W-2	7	1	8
S-2, W-3	3	4	7
N-1, W-1	1	5	6
S-2, W-1	3	3	6
S-3, W-1	3	3	5
S-1, E-1	2	3	5
N-0, E-1	1	4	5
S-1, E-0	4	1	5
S-2, E-0	1	4	5
S-3, W-3	5	0	5
S-2, W-2	3	1	4
N-0, W-2	2	1	3
S-1, W-1	3	0	3
N-1, E-1	1	1	2
N-1, E-0	0	1	1
N-1, W-2	1	0	1
N-2, W-1	1	0	1
S-1, W-3	1	0	1
S-3, E-0	0	0	0
N-0, W-3	0	0	0
S-3, E-1	0	0	0
N-2, W-2	0	0	0
N-3, W-2	0	0	0
N-3, W-1	0	0	0
N-2, E-1	0	0	0
N-2, E-0	0	0	0
N-2, W-3	0	0	0
N-3, E-0	0	0	0
N-3, W-3	0	0	0

$p < .05$) and seasonal items all together ($r_s = .96$, $p = .001$). In the Old Forest there was also a positive correlation between distance traveled and fig fruit availability ($r_s = .62$, $p < .05$, one-tailed test, $n = 8$), which is consistent with the positive correlation that was found between fig-eating and fig availability. It would appear that when figs are more available in the Old Forest the monkeys tend to eat more and travel more to get them.

On the other hand, in the Lutz Ravine, the correlation between travel distance and fig availability was negative ($r_s = -.47$, $p = >.10$, one-tailed test, $n = 9$). It may be that when figs are more available in the Lutz Ravine the monkeys travel less between patches, having a number of fruiting figs all in the same area. This was certainly the case in the July sample, the month in which the troop traveled the shortest mean daily distance; at that time two *Ficus insipida* trees (T-36 and T-37), which are less than 35 meters apart, were both fruiting and were used heavily (see table 5.1). Also, in November, the month in which the troop traveled the next short-est daily distance, there were at least three *F. yoponensis* trees all fruiting in the same area and two were used heavily.

Significant positive correlations were found between the number of species eaten daily and time spent traveling and distance trav-eled, but not travel rate ($r_s = .2977$, $p = .006$; $r_s = .327$, $p = .004$; $r_s = .09$, $p = .39$; $n = 85$). This suggests that howlers traveled more and further when their diets were more diversified in terms of species. In chapter 4 it was shown that howlers had more diversi-fied diets at times when there were a greater number and variety of seasonal food sources available. So it would appear the pressures to minimize travel were less strong at times when there were a greater number and variety of foods available.

Supplying Area

SIZE.   A gross estimate of the size of the supplying area for each troop was made by combining on a map all of the daily ranging pat-terns over the sample period and multiplying the distance between the extreme east and west points times the distance between the extreme north and south points (see figures 5.1 and 5.2). For the Old Forest this gave an area of 39.23 hectares; for the Lutz Ravine, 43.73 hectares.

Perhaps a better measure is the area of a convex polygon that en-closes all sightings of the troop during the sample period (Waser and Floody 1974). For the Old Forest this gave an area of 31.66 hectares; for the Lutz Ravine, 31.09 hectares (see figures 5.3 and 5.4).

As pointed out by Milton and May (1976), the size of a supplying

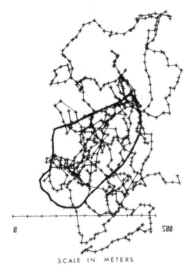

SCALE IN METERS

*Figure 5.1. Combined Daily Travel Patterns for Total Sample Period in
the Old Forest Study Area
Heavy lines indicate footpaths.*

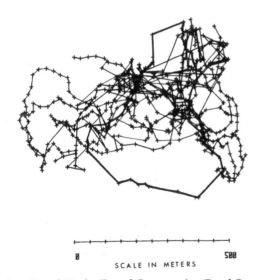

SCALE IN METERS

*Figure 5.2. Combined Daily Travel Patterns for Total Sample Period in
the Lutz Ravine Study Area
Heavy lines indicate footpaths.*

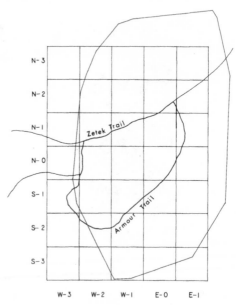

Figure 5.3.   Supplying Area for the Old Forest Study Troop
The inner line is the trail network (footpaths). The outer line encloses all
sightings of the study troop during the total sample period in a convex
polygon which indicates the total supplying area = home range. Each
quadrat = 1 hectare.

area can be related to body weight, diet, and troop size as well as
features of the habitat. Since these factors were very similar between the two areas, it is not surprising that the sizes of the supplying areas were similar. The main difference between the two
areas—the much greater number of *Ficus* trees in the Lutz Ravine—
did not seem to affect the size of this area. Though howlers in the
Lutz Ravine ate more fig products than howlers in the Old Forest,
the number of species eaten in each area was almost identical. Thus
apparently the supplying area necessary to provide the dietary mix
required by howlers was around thirty-one hectares for each troop.

It can be seen that the supplying area for each troop could be
traversed in one day's travel (i.e., in 1.5 hours of time spent traveling). The size of the supplying area is approximately equal to the
square of the distance traveled by howlers in that amount of time.

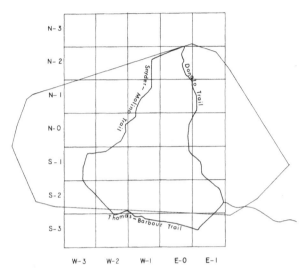

Figure 5.4.   Supplying Area for the Lutz Ravine Study Troop
The inner line is the trail network (footpaths). The outer line encloses all
sightings of the study troop during the total sample period in a convex poly-
gon which indicates the total supplying area = home range. Each quad-
rat = 1 hectare.

It may be that for howlers, the size of the supplying area is limited
by the area that can be efficiently monitored.

In this regard, there is evidence that the size of howler troops on
Barro Colorado may be limited, possibly by factors related to the
size of the supplying area. Carpenter's (1934) census of 1932 showed
twenty-three howler troops on Barro Colorado and his census of
1958 (1962) showed forty-four troops. In his first census the mean
troop size was 17.3 (±6.9) and in his last census it was 18.5 (±9.4).
Thus while the overall population increased, troop size tended to
remain the same and new troops were apparently being formed by
split-offs from old ones. Further, in observing a large troop in the
Fairchild area of Barro Colorado (i.e., 24 to 27 animals), I noted
that the troop often split up for foraging and remained apart for
long periods during the day. This was an extremely rare occurrence
in my two study troops, which tended to remain together in the
same tree almost continuously when they were not traveling. It

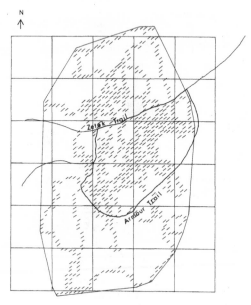

Figure 5.5.   Area Used by Old Forest Study Troop During the Total Sample Period.
Hash marks show each 12.75 × 12.75 meter quadrat entered by the troop.

would therefore appear that for howlers on Barro Colorado there is an optimal troop size, beyond which there are pressures for animals to split off and form new troops. Since the larger the troop size, the larger the required supplying area is likely to be (Milton and May 1976), these pressures may come from a decrease in the net energy return per animal from foraging over a large area.

DIFFERENTIAL USE.   For each study troop the portion of the supplying area that was actually used during the sample period was estimated by counting the number of 12.75m × 12.75m quadrats entered by the troop and summing the areas of these quadrats (Waser and Floody 1974). By this measure, the study troop in the Old Forest used 10.92 hectares of its supplying area and the study troop in the Lutz Ravine used 14.31 hectares (see figures 5.5 and 5.6). In other words, the Old Forest troop actually used about 34 percent of its supplying area while the Lutz Ravine troop used about 46 percent. Thus, both troops appear to have been selective in their use of the supplying area.

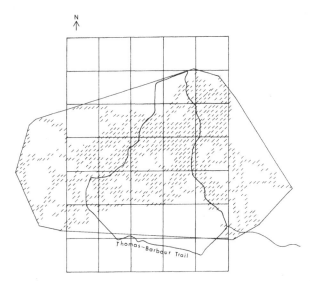

Figure 5.6.   Area Used by Lutz Ravine Study Troop During the Total Sample Period.
Hash marks show each 12.75 × 12.75 meter quadrat entered by the troop.

An analysis of the travel routes showed that certain areas were used more heavily than others, simply in terms of the number of routes they contain (see figures 5.7 and 5.8). Further, an analysis of the location of primary food trees showed that some areas contained more of such trees and were visited more often for feeding bouts than other areas (see table 5.5). In the Old Forest these areas were the same as those used most heavily for travel, except for the N–1, W–1 hectare but, as figure 5.9 shows, this area is between two major feeding areas and was crossed by the troop going from one such area to the other. Similarly, in the Lutz Ravine, the areas used most for feeding were the same as those used most for travel, except for the N–0, W–1 hectare, which again is between two major feeding areas (see figure 5.10). Thus, in both study areas the selective use of supplying area was related to feeding.

When the areas used most for feeding were examined for potential food sources, it was found that these areas tended to have higher densities of particular species preferred by howlers than areas that were used less. For example, in the Old Forest the N–2, E–0 hectare has two *Ficus yoponensis* and four *F. insipida*—a very high density,

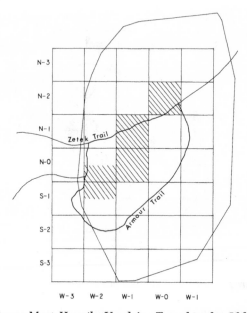

Figure 5.7. Areas Most Heavily Used for Travel in the Old Forest Study Area
Shaded quadrats had the heaviest concentration of travel routes. Each quadrat = 1 hectare.

especially in relation to the total number (21) of such trees in the 35 hectares of the Old Forest study area. The N–0, W–2 hectare has a much used *F. yoponensis*, plus a *Brosimum alicastrum* and two huge *Anacardium excelsum*. The N–1, W–2 hectare has two *F. yoponensis* and one *F. insipida* plus a *Platypodium elegans*. Since data on all potential food trees are available only for the three sample plots in the Old Forest (N–2, E–0; N–0, W–1; and S–3, W–1), detailed comparisons cannot be made between the various hectare quadrats except for *Ficus* trees, all of which were located to collect phenological data. But it is evident that three of the four most heavily used hectares have more *F. yoponensis* and/or *F. insipida* than the mean number per hectare in the study area (0.60).

In the Lutz Ravine the hectares used most heavily for feeding had high densities of *Ficus yoponensis* and *F. insipida* (see table 5.4). In fact, when the 15 hectares that form the central part of the

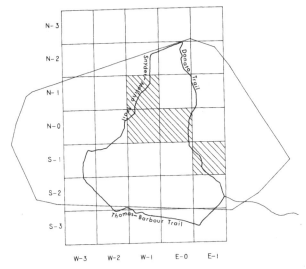

*Figure 5.8.   Areas Most Heavily Used for Travel in the Lutz Ravine Study
Area*
Shaded quadrats had the heaviest concentration of travel routes. Each
quadrat = 1 hectare.

study troop's supplying area were examined, it was found that the
7 most heavily used hectares had significantly higher densities of
these two species than the other 8 hectares (M–W, p = .0125). The
mean number of individuals of the two species per hectare was 6.43
in the most heavily used quadrats versus 2.13 in the others. Thus,
with respect to *F. yoponensis* and *F. insipida*, howlers in both study
areas appear to have selectively used areas where they had a higher
probability of finding food.

My subjective impression is that these areas also have relatively
high densities of some other food species. For example, in the T-36
area there are not only many fig trees, but there are also two groups
of *Spondias mombin* (fruit source), two *Maquira costaricana* (fruit
and leaves), two *Platypodium elegans* (leaves and flowers), a *Poulsenia armata* (leaves), and a *Luehea seemannii* (leaves). Thus it is
an area with a good number and variety of preferred food sources.

This raises the question of whether howlers might be affecting
the structure of the forest in their favor. Several researchers have

TABLE 5.5. Visits to Primary Food Trees

OLD FOREST		
Hectares	No. of Food Trees	No. of Visits
N-0, W-2 (south) S-1, W-2 (north)	4	27
N-2, E-0	9	19
N-1, W-2	8	13
N-0, W-1	8	10
S-2, E-0	*Cecropia* grove	6
N-3, E-0	2	5
N-1, E-1	3	4
S-2, W-1	3	4
S-3, W-1	3	4
N-2, E-1	2	2
N-0, E-1	1	1
S-3, W-2	1	1
N-1, E-0	1	1
N-3, E-0	1	1

LUTZ RAVINE		
Hectares	No. of Food Trees	No. of Visits
N-0, E-0	14	47
N-1, W-1	7	37
S-1, E-1	9	31
N-0, W-2	7	29
N-1, W-2	3	15
N-2, E-0	5	13
N-0, W-1	6	12
S-1, W-2	5	10
N-0, E-1	4	8
S-2, W-3	3	7
S-1, E-0	2	4
N-2, E-1	2	4
N-1, E-0	3	3
S-3, W-3	1	1
N-1, E-1	1	1
S-1, W-1	1	1

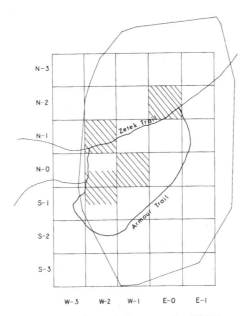

Figure 5.9.   Visits to Principal Food Trees in the Old Forest Study Area Shaded quadrats had the highest number of feeding visits. Each quadrat = 1 hectare.

discussed fruit seed dispersal by means of the howler gut (Hladik and Hladik 1969; Hladik et al. 1971). Howlers typically enter a fruiting tree and take one or more meals from it, often swallowing the fruit seeds. They then often travel to another area to eat foliage and sleep. The defecation pattern of howlers is very predictable. Howlers generally defecate before beginning their early morning travel to food trees and again after prolonged periods of rest. Thus about half of the time, seeds from fruiting trees would tend to be dispersed in areas near leaf sources, since these are usually eaten late in the day. It would seem that over a long period of time howlers might shape the forest to some extent, but chance would take a heavy toll here, since the number of seeds that germinate and survive out of the virtually millions that are produced per species per annum in a tropical forest must be extremely low. Still, in time, clusters of howler food trees (fruit and leaf) might be found to be

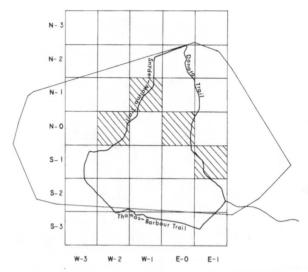

Figure 5.10. Visits to Principal Food Trees in the Lutz Ravine Study Area Shaded quadrats had the highest number of feeding visits. Each quadrat = 1 hectare.

in significantly closer proximity to one another than species not eaten by howlers.

EXCLUSIVITY. The supplying area of a group of animals can be defended from encroachment by conspecifics. In such cases, this area is typically referred to as a "territory" (Burt 1943). A stricter definition of territory (and the one used hereafter) requires that the area be *successfully* defended (Brown and Orians 1970). In other cases, the supplying area may be used by one or more conspecific groups with varying degrees of overlap. In such cases the area is generally referred to as a "home range" (Jewell 1966). If certain areas in a home range are used exclusively or very heavily by one group, these are generally referred to as "core areas" (Jay 1965).

Howler monkeys, according to the above definition, are not territorial. From one to four different howler troops used various parts of the home range of my study troop in both study areas, and in the Lutz Ravine there seemed to be about 100 percent overlap between my study troop and one other troop—i.e., they were using the same supplying area, though the other troop did appear to use a fringe area on the northwest more than the study troop. The selective pressures for territoriality are many and complex and have been the subject of much discussion both in the primate literature and elsewhere (e.g., Burt 1943; Ellefson 1968; Brown and Orians 1970). The relevant question here is whether or not the defense of a territory—an exclusive area—would be advantageous as part of the howler foraging strategy.

In terms of the theory of foraging strategy, the defense of a territory would be advantageous if such behavior resulted in a net energy benefit to the animal (Brown 1964; Brown and Orians 1970; Schoener 1971). In other words, the cost of patrolling and defending the territory would have to be outweighed by the benefit from having exclusive use of food resources—that is, exclusive with regard to conspecifics. As noted, howler food sources are very patchy, in both space and time. In chapter 4 it was shown that even at times when fruit (the howlers' main source of ready energy) was most available, howlers ate fruit only up to a point and then ate leaves, presumably to get the other nutrients they required. Thus at times there was apparently more than enough fruit available in a given area to support a howler troop—a conclusion also reached by Hladik et al. (1969; 1971). At such times, a howler troop would get little or no benefit from defending a particular fruit tree, while such behavior would be costly both in terms of energy expenditure and in loss of opportunity to get the other required nutrients.

On the other hand, it has been shown that when the overall supply of fruit was very low, howlers ate more leaves, including more mature leaves, exploited available food sources more intensively, and traveled less. There appeared to be more inter-troop competition for food sources, but no more evidence of territorial behavior—i.e., patrol and defense of an exclusive area. Presumably this was because there would not have been sufficient rewards for the energy spent in such behavior. The pressures to minimize travel would have then been stronger.

Howlers do have a behavioral feature—the dawn chorus—which is believed to function as an inter-troop spacing mechanism (Carpenter, 1934; Chivers 1969; and others). Howlers often give the dawn chorus, feed, rest, and then set out on a travel bout through the forest to a new area. While traveling, or directly on entering this new area, the males of the troop will often give a series of howls, which I believe function to advertise the fact that they have changed locales. As a general rule, howler troops tend to avoid one another and always show strong antipathy toward strange conspecifics. Inter-troop encounters are costly and dangerous. Time and energy are expended running, howling and even fighting; members become separated from their troops; estrous females from one troop may come into contact with extra-troop males. For all of these reasons, it is not advantageous for one howler troop to encounter another and thus it pays to use the howling vocalization to avoid one another.

When two troops do encounter each other, the outcome appears to be determined by an inter-troop dominance hierarchy. My Lutz Ravine troop appeared dominant over all other troops that overlapped with it and was never noted to give way or retreat in inter-troop encounters. The data are ambiguous on my Old Forest study troop. Success in an inter-troop encounter appears to depend on the unity and fighting abilities of the coalition of adult males of particular troops. These animals often move out in a group from their females and young and physically drive away the males of an opposing troop, who flee through the trees, leaving their females and young to follow after them as best they can. When two strong troops meet there may be a long series of vocal battles and even some physical scuffling before the conflict is resolved. Then, sedately, the repulsed troop will alter its course and gradually withdraw from the area. If there is an inter-troop dominance hierarchy, it could be a means of allocating food resources at times of low overall supply, ensuring that at least some animals get enough to eat (Murton et al. 1966).

SUMMARY

Because of the relatively low energetic content of much of their food, I expected that howlers would maximize their net energy return from foraging by minimizing travel. Because of the patchy distribution of their food in both space and time, I expected that howlers would be goal-directed in their travel and very selective in their use of supplying area. My data indicate that:

1. Howlers on Barro Colorado traveled almost solely to get to food sources.

2. They traveled in single file, which would not be the most efficient way of covering an area if they had been randomly searching for food; but it would be the most efficient way of traveling to a goal.

3. Travel patterns were oriented around "pivotal" trees, which were always primary food sources and were used repeatedly during a given day or period of several days. Such trees, which were usually fruit sources, were apparently used as bases from which the troop traveled to various leaf sources, having a known source of ready energy to return to; and while foraging in the area, the monkeys could monitor other potential fruit sources.

4. The troop often returned to pivotal trees by the same routes and, further, they repeatedly used the same routes ("arboreal pathways") to get from one major patch of food to another. These pathways may have been used to help locate foods, since they led to individual trees that were used repeatedly as food sources or to areas where there were patches of such trees. The monkeys visited these trees frequently and regularly, which suggests that they were monitoring the trees as potential food sources.

5. The supplying areas used by both study troops were approximately the same size. These areas could be traversed in a little more than an average day's travel (i.e., in 1.5 hours). Thus, all potential food sources within the area could be reached with a limited expenditure of energy in travel.

6. Certain parts of the supplying area were used more heavily than others. These were areas where the densities of particular food species (e.g., *Ficus yoponensis* and *F. insipida*) tended to be higher than in other areas, which indicates that the monkeys were being selective in their use of the supplying area. This would also reduce the area being monitored for food sources and thereby presumably increase the efficiency of such behavior.

7. There was considerable overlap in the supplying areas of adja-

cent troops. Howling was apparently used as an inter-troop spacing mechanism. Troops of howlers seemed to avoid contact with one another, and when such encounters did occur the outcome seemed to be determined by inter-troop dominance hierarchies.

All of the above behavioral features would have the effect of minimizing travel and thereby minimizing the costs of foraging.

6 TIME SPENT FORAGING

IN MOST models of foraging strategy (MacArthur and Pianka 1966; Emlen 1966; Schoener 1971; and others), the efficiency of foraging is measured by the net energy yield/foraging time. The underlying assumption is that the amount of time spent foraging is indicative of the cost of foraging, both directly in the sense of energy cost and indirectly in the sense of opportunity loss (i.e., time spent foraging cannot be spent on other activities such as reproduction). There is also the risk of an animal being more exposed to predators while foraging. It is therefore relevant to examine the time spent foraging.

There are obviously times in an annual cycle when the foods preferred by howlers on Barro Colorado are more abundant than at other times. If howlers ate the same proportions of such foods (young leaves, fruits, and flowers) throughout an annual cycle, I would have expected them to spend more time foraging when these foods were less abundant. However, at such times howlers ate less fruit and more leaves, including more mature leaves. Thus when food was less abundant, howlers were being less selective in their food choices, as predicted by the models of MacArthur and Pianka (1966), Emlen (1966), and Schoener (1971). This would tend to reduce the time spent traveling, especially since leaves are a less patchy resource than fruit. Since leaves also provide considerably less ready energy than fruit, there is less incentive to spend time foraging for leaves. For these reasons, having found that my study troops were more folivorous at times when seasonal foods were less abundant, I expected that they would spend less time foraging during such periods.

METHODOLOGY

Again, the activities of all visible members of my study troop were noted at five-minute intervals from dawn to dusk on each sample day. The percent of time spent that day in each activity was then calculated as it was for feeding. Since I was investigating foraging behavior, I was interested in the time spent in three main activities: feeding, traveling, and resting. Since travel was almost always related to feeding, the time spent feeding and traveling, together, gives a good indication of the time spent foraging; and the time spent resting gives an indication of the amount of time the monkeys had left after meeting their nutritional requirements. The data were analyzed on a daily basis, as explained in chapter 4.

RESULTS

Overall

Combined data from the two study areas showed that howler monkeys spent 16.24 percent of their time feeding, 10.23 percent traveling, and 65.54 percent resting (table 6.1). These three activities thus accounted for 92 percent of a typical howler day.

By far the most time was spent resting. It is interesting to see that feeding and traveling together accounted for only 26.47 percent of

TABLE 6.1. Mean Percent of Daily Time Spent in Principal Activities

	Wet	Transition	Dry	Overall
Combined				
Feeding	15.51	20.77	15.90	16.24 ± 4.32
Traveling	9.72	10.81	10.77	10.23 ± 3.65
Resting	65.89	62.02	66.23	65.54 ± 6.98
Old Forest				
Feeding	15.16	19.66	16.79	16.23 ± 4.05
Traveling	9.15	10.38	11.60	10.19 ± 3.88
Resting	66.47	63.58	64.66	65.45 ± 7.18
Lutz Ravine				
Feeding	15.82	21.88	15.07	16.24 ± 4.59
Traveling	10.22	11.24	10.00	10.26 ± 3.47
Resting	65.38	60.46	67.70	65.51 ± 6.80

total time. Thus, despite the patchiness of their food resources, the monkeys were able to meet their nutritional requirements with only a little more than three hours of foraging each day.

Spatial Differences

There were no significant differences between the two study troops in time spent feeding, traveling, and resting (table 6.1). Indeed, the overall results were remarkably similar despite the differences in diet.

Temporal Differences

Combined data from the two study areas show that more time was spent feeding in the transition season than in the wet or dry seasons (K-W, $p = .003$, $N = 85$). This was the pattern in both areas, though the difference was statistically significant only in the Lutz Ravine ($p = .001$, $N = 45$). The increase in feeding time during the transition season may be related to the increase in the proportion of leaves in the diet. Generally, it seemed to take the monkeys longer to become satiated when eating leaves than when eating fruit. There was a positive correlation between time spent eating leaves and time spent feeding ($r_s = .1277$, $p = .086$, $n = 85$) and a negative correlation between time spent eating fruit and time spent feeding ($r_s = -.1008$, $p = .178$, $p = 85$), though neither of these was statistically significant.

The similarities between areas and between seasons in the time spent foraging were more remarkable than the differences. Throughout the annual cycle the amount of time spent foraging was relatively low, generally from two to four hours per day, and it fluctuated very little despite the seasonal changes in food availability. There were no significant correlations between the time spent foraging and the availability of young leaves, fruit, flowers, or all seasonal foods. Thus it would appear that by shifting their dietary focus, the monkeys avoided spending much time foraging when preferred foods were less available.

It may be adaptive for howlers to spend a regular, relatively low, percent of time foraging throughout the annual cycle. When energy-rich foods are abundant, howlers eat more of such foods, yet they remain relatively inactive; in this way, they may be able to build up reserves of energy. When energy-rich foods are scarce, they eat more leaves, avoiding the greater expenditure of time and energy they would have to make to maintain the same proportion of fruit in their diet and possibly living to some extent on reserves of energy accumulated during times when fruit is abundant. As noted earlier, if an animal spends less time foraging, it requires less energy; and, in this respect, howlers appear to have evolved a low-energy way of life, i.e., they are energy-conservers, as indicated by the regular, relatively high percentage of time they spend resting (65.54 percent ± 6.98).

SUMMARY

1. During the sample period, howlers spent a mean of 26.47 percent of the day foraging (traveling and feeding), or 3.18 hours. The mean percent of time spent foraging was almost exactly the same for both study troops, despite differences in diet and habitat.

2. More time was spent feeding during the transition season than in the wet and dry seasons. This may be related to the greater proportion of leaves in the diet during the transition season, the time of lowest overall production of seasonal foods. There were, however, no seasonal differences in time spent traveling.

3. Throughout my study period the percent of time which howlers spent foraging was remarkably low and regular, despite the seasonal changes in food availability. At times when fruit was abundant they had a diet in which fruit and leaves were about equally important. When fruit was scarce, they did not spend more time searching for it; rather, they ate mainly leaves, which should have been easier to find. Thus, by shifting their dietary focus they continued to spend a relatively low percentage of time foraging.

4. The monkeys spent by far the greatest percent of the day resting (65.54 percent ± 6.98). This suggests that energy conservation is an important part of their foraging strategy.

7 ASSESSING THE HOWLER FORAGING STRATEGY

THE CLASSICAL theory of foraging strategy postulates that the net return of energy from time or energy spent in foraging will be maximized by natural selection (MacArthur and Pianka 1966; Emlen 1966; Schoener 1969, 1971; and others). A few researchers have pointed out the limitations of energy models in dealing with the problems of primary consumers. Westoby (1974) proposed that the objective of the foraging strategy of a large generalist herbivore is to optimize the nutrient mix within a given total bulk of food, rather than to maximize the energy yield/foraging time. He concluded that since the foraging activities of herbivores are limited by digestion time, rather than by search or pursuit time, food *quality* will be more important than availability in determining the choice of diet. Thus diet selection would be constrained by availability of foods only at rather low levels (see, for example, Sinclair 1974, 1977).

Freeland and Janzen (1974), who also considered the problems of herbivores, suggested that choice of diet may be affected by the presence of secondary compounds in most plant parts (see also McKey 1975, 1978). Since many of these compounds impede protein digestion and can be toxic, herbivores must have mechanisms for counteracting their effects or for degrading and excreting them— i.e., microsomal enzymes and/or gut flora (Schuster 1964; Scheline 1968; Williams 1969, 1971; and many others). The limitations of these mechanisms, however, may force such animals to eat foods high in nutrients to counterbalance losses incurred by digestibility-reducing components or to eat small amounts of several kinds of food to avoid a toxic overload. Freeland and Janzen (1974) therefore predict that herbivores should have a searching strategy and/or body size adapted to optimizing the number of foods available with respect to the amounts of each that can be eaten, rather than maximizing the availability of any given food. They also predict that once an herbivore has established a range of food species and parts that it can consume with impunity, it should continue to feed on them for as long as possible.

What is stressed by these latter researchers is the importance to a primary consumer of selecting foods that contain the specific nutrients which they require and avoiding or limiting the intake of foods that are low in nutrients and/or contain secondary compounds which they cannot degrade efficiently. Thus, for a primary consumer the problem of food selection is much more complex than the one described by the classical models. The other main factors that should be considered by a theory of foraging strategy for primary consumers are the efficiency of food digestion, the distribution of food sources, and the costs of food procurement. The objective of the foraging strategy of a primary consumer should be an optimum mix of energy and nutrients within a given total bulk of food at a minimum cost of procurement. The relationship between the above factors is obviously very complex. But each of these factors can be considered separately to explain the possible adaptive function of certain behavioral and morphological features that have emerged from the data presented in the preceding chapters.

Food Digestion

Old World primate folivores—colobines, indriids, lepilemur— have digestive specializations that apparently enable them to use cellulose and hemicelluloses as sources of energy. The greatly enlarged and elongated sections of their digestive tracts provide an extensive field for the activities of bacterial flora, which break down structural carbohydrates through fermentation and release compounds (i.e., volatile fatty acids) that are believed to make an important contribution to the energy metabolism of the host animal (Bauchop 1971). The bacteria also synthesize protein, which could provide additional high quality protein and/or certain important amino acids to the host animal (Moir 1965; Blackburn 1965). Further, it has been shown that some types of gut flora perform detoxicatory functions, which would reduce demands on the microsomal enzymes of such animals (Schuster 1964; Scheline 1968; Williams 1971). These animals should therefore be able to utilize foods that are high in structural materials (and certain secondary compounds) more efficiently than animals lacking such specializations. This would, in turn, reduce their costs of food procurement, since such foods (e.g., mature leaves) are very abundant and always available in a tropical forest. It would also reduce the pressure to select foods so as to get a balance of energy and nutrients, since the activities of the gut flora could provide both ready energy and high quality protein.

Even such animals, however, should be subject to the pressure to select foods with the highest energetic and nutritive content (and lowest toxic content) relative to the costs of procurement. Therefore, at times when seasonal foods (young leaves, fruits and flowers) are relatively abundant, they should eat such foods. At times when seasonal foods are relatively scarce and therefore the costs of procuring them are relatively high, such animals should eat more perennial and abundant foods such as mature leaves and bark. The available data on the foraging behavior of Old World specialized primate folivores seem to conform with this expectation (e.g., Ripley 1970; Hladik and Hladik 1972; Clutton-Brock 1975; and others).

Material presented in chapter 1 indicates that howler monkeys lack the sacculated stomach or greatly enlarged hindgut characteristic of many Old World leaf-eating primates. However, as noted, this does not preclude the possibility of fermentation activities in howlers. Various monogastrics with relatively simple hindguts are known to ferment plant structural carbohydrates (e.g., rats, swine, humans; Keys et al. 1969; Spiller and Amen 1975).

Toward the end of this field study, having noted the extremely strong dependence of howlers on seasonal rather than perennial foods, I wanted to determine what benefits, if any, they might obtain from end products of fermentation. Thus I carried out a set of feeding experiments on three wild howlers captured for these experiments with tranquilizing darts. The animals were fed two different diets—one made up primarily (81 percent) of fig fruit and the

other made up primarily (87 percent) of fig leaves (see Nagy and
Milton 1979a and Milton et al. in press, for details of experimental
techniques). Results showed that considerable cellulose and hemicelluloses disappeared in transit on these diets. On the fruit diet,
which was high in indigestible fig seeds, an average of 23 percent
of the cell wall material disappeared in transit and on the leaf diet
41 percent. In addition, howlers showed high proportions of metabolic fecal nitrogen. From the amount of structural material degraded, it appears that howlers may obtain from one-fourth to one-
third of their required daily energy (estimated at 355 kJ per kg. per
day; Nagy and Milton 1979b; doubly-labeled water technique) from
fermentation end products (i.e., volatile fatty acids), depending on
the types and amount of foods eaten. Unlike some digestively specialized Old World folivores, however, howler monkeys apparently
cannot rely on diets composed entirely or almost entirely of high-
fiber perennial foods such as mature leaves. Even with the benefits
provided by fermentation end products such a diet is apparently
too low in quality to sustain the less-specialized howler. No field
study to date has shown that howlers are strongly dependent on
mature leaves in the diet at any time of year. (Should howlers elsewhere be shown to eat considerable amounts of mature leaves in
some future study, it is hypothesized that such leaves will be considerably higher in quality than most mature leaves analyzed from
the Barro Colorado forest.) Further, captive howlers consistently
refused to eat most mature leaves offered them during the feeding
trials even though at times they were obviously quite hungry. It
thus appears that howlers do not fulfill their daily requirements for
energy and other nutrients primarily through end products of fermentation. This should increase the pressure on them to select foods
so as to get a balance of energy and other nutrients, which in turn
should increase the costs of food procurement.

It would also increase the pressure to select foods with the highest energetic and nutritional content (and lowest toxic content) relative to procurement costs. In a tropical forest, however, higher quality seasonal foods are less abundant and less available than mature
leaves; in the Barro Colorado forest such foods are very patchy in
both space and time. Thus, there should be a strong pressure on

foraging behavior to minimize the costs and risks of procuring such foods. Behavioral features that advance this objective should be favored by natural selection.

Given the apparent limits of their digestive system, the foraging strategy of howlers should be to select foods of a relatively high quality, to minimize the costs of procuring such foods and generally to conserve energy, especially at times when the supply of preferred foods is low.

Food Selection

Howlers on Barro Colorado spent an overwhelming percent of feeding time on seasonal items (young leaves, fruits, and flowers) even though such items were less abundant and less available than perennial items. Thus, as predicted by Westoby (1974), food quality was evidently more important in determining the choice of diet than availability. The monkeys did, however, eat some mature leaves at times when seasonal foods were scarce, which indicates that they were constrained by food availability at low levels.

Typically, they ate both leaves and fruit each day. When fruit was more abundant they ate more fruit, but only up to a point, as they continued to maintain a substantial proportion of leaves in their diet. This indicates that they selected food categories so as to get a balance of energy and nutrients (i.e., protein from leaves and nonstructural carbohydrates from fruit). When fruit was scarce they ate more leaves, which are less patchy than fruit and should therefore be easier to procure. Thus, when the cost of procuring fruit was unacceptably high, they became more folivorous, though even then they continued to be selective and eat primarily young leaves. At such times, their diet must have provided considerably less ready energy, but they traveled less and therefore must have expended less energy in food procurement. They may also have had some energy reserves, accumulated during times of fruit abundance.

The percent of time howlers spent feeding on particular species was not closely related to the density of these species in the Barro Colorado forest, as indicated by sample plots, or to the number of months in which these species produced seasonal items, as indicated by phenological data. Further, they did not feed at all on many species, some of which had relatively high densities. Thus, some other factor—apparently food quality—was more important than availability in determining the species used as food sources.

When eating leaves, howlers generally ate small amounts from several species, as predicted by Freeland and Janzen (1974). However, there are not enough data to show that they were mixing leaf sources so as to minimize the costs of detoxication (though they did tend to avoid leaves from certain families with known toxic properties, e.g., *Anacardiaceae*). Indeed howlers may have eaten leaves from several species each day simply because the supply of young leaves on any one tree was not sufficient to meet the protein requirements of a howler troop. But in either case they were being selective, since there generally would have been sufficient leaves on only one tree for an entire howler troop if animals had eaten both young and mature leaves.

Food Procurement

For any animal the pressure to be selective in its feeding behavior is opposed by the pressure to control the costs of food procurement. Thus, in a habitat where there is more than one potential food source, the more selective an animal is, the higher its costs of procurement are likely to be. The cost problem could be especially great for an animal that is selecting foods which are very patchy in both space and time. As this is the situation of howlers on Barro Colorado, they should have developed behavioral and morphological features to minimize the costs of procuring such foods. The features that emerge from the data follow.

DIVERSIFICATION.   In selecting seasonal items, howlers on Barro Colorado used a great number of different plant species as food sources. They used 109 different species during the sample period and 7 to 8 each day. Since most of the potential food sources in the Barro Colorado forest have clumped distributions and/or relatively low densities, an increase in the number of species used as food sources would reduce the probable distance howlers would have to travel from one potential food source to the next. Thus, by diversifying their sources of seasonal foods, howlers on Barro Colorado reduced the probable costs of food procurement. This strategy, however, must have been constrained by the physical, nutritional and toxic properties of potential foods, since many plant species and parts were not used as food sources, as noted above.

FLEXIBILITY. Howlers on Barro Colorado changed the species used as food sources from day to day and from month to month. Changing species from day to day would be advantageous if the cost of returning to a tree used the previous day or finding another tree of the same species were not justified by the energy and nutrients offered by such a food source. As noted earlier, young leaves were less patchy than fruit and were generally less abundant in particular trees, so that both the pressure and the incentive to search for particular leaf sources should have been less. The daily turnover of leaf species should therefore have been higher than that of fruit species, and it was (62 percent vs. 35 percent). This suggests that changing species from day to day increased the efficiency of foraging. Leaf species might also have been changed to obtain the best complement of nutrients, minimize effects of digestibility-reducing components, or avoid an overload of particular toxins. If so, this would have increased the efficiency of foraging by reducing the costs of food digestion.

By changing species from month to month, howlers took advantage of the different phenological strategies of the plants in their habitat and were able to eat mainly seasonal foods throughout an annual cycle, even though such items were available from particular species only during short periods in most cases. Thus, they maintained a higher quality of diet than if they had continued to eat both seasonal and perennial foods from the same species throughout the year.

SPECIALIZATION. Though howlers on Barro Colorado used a great number of food sources, at the same time they specialized to some extent on certain plant species and families. Trees of these species and families were generally used as sources of more than one category of food and were therefore used at more different times of an annual cycle than if they had been used for only one category, since most individual trees produce new leaves, fruit and flowers at different times. In this sense, the potential supply of food in such trees was less discontinuous than in trees that were used for only one category—with the exception of a few continuous flushers. As noted above, there are not enough data in all cases to explain why howlers ate foods from some species and ignored potential foods from others, but in general it appears that the seasonal items from species they exploited for more than one food category had properties, both physical and nutritional, that made them particularly profitable to exploit; and by using such species more heavily than others, howlers should have reduced the costs of food procurement, since the probability of finding food would have been higher than in trees of species which, for some reason, they could exploit for only one food category.

Howlers on Barro Colorado made especially heavy use of two species of the genus *Ficus*. They used trees of these species as sources of both young leaves and fruits, and used them throughout the annual cycle. As explained, individuals of these two species are asynchronous in their phenological cycles and therefore, where there are enough of them, may provide a continuous supply of seasonal foods. This is the case in the Lutz Ravine, where there is an unusually large number of such trees. The Lutz Ravine study troop spent almost half of total feeding time on these two species and ate fig products on all but one of the sample days. Thus, where there were continuous sources of seasonal foods, howlers made use of them and thereby reduced the costs of food procurement. The pressures to diversify and be flexible were evidently less strong in the Lutz Ravine than in the Old Forest where there were considerably fewer fig trees; the diet of the Lutz Ravine troop was less diverse and less variable. Since the Lutz Ravine is a disturbed area, the behavior of the Old Forest study troop is probably more representative, but the behavior of my Lutz Ravine study troop helps to elucidate the procurement problem and shows how, in an area where there are continuous sources of seasonal foods, howlers have adapted by specializing on these sources to a great extent.

As noted, Freeland and Janzen (1974) predicted that, because of the problems that might be caused by the secondary compounds in plant foods, once an animal has established a range of food species and parts that it can eat with impunity, it should continue to feed on them for as long as possible. The strong preferences shown by howlers on Barro Colorado for foods from certain species and families appear to conform with this prediction, which supports the hypothesis that they have developed mechanisms for efficiently degrading particular sets of compounds in the foods they eat from these species and families. More data are needed on the secondary compounds of potential plant foods and on the detoxication system of howlers to further test the Freeland-Janzen hypothesis.

GOAL-DIRECTED TRAVEL. It can be shown that where food sources are patchy in both space and time, goal-directed travel would be a more efficient strategy for food location than uniform or random travel. If a particular food is patchy, then the probability of finding it is higher in some places (and at some times) than others. So if the animals were in an area where there was no food, neither uniform nor random travel would necessarily increase the probability of finding some; and if the animals were in a place where there was food, uniform or random travel might actually decrease the probability of finding more (i.e., by taking the animals out of the patch). Thus, if howlers on Barro Colorado are minimizing the costs of food procurement, they should have a strategy of goal-directed travel.

The data on ranging behavior showed that howlers on Barro Colorado traveled almost solely to get to food sources, as they had no particular sleeping sites and no apparent need to find water. They usually traveled in single file, which would not be the most efficient way to search for food if they had been traveling at random; it would, however, be the most efficient—and the safest—way to travel to a goal. Further, their ranging patterns were oriented around pivotal trees, which were always primary food sources and which they used repeatedly during the same day and/or during a period of several or more days. In returning to such trees, they often used the same routes. All of this evidence suggests that travel was goal-directed.

This raises the question as to how the monkeys knew where and when to go for food. As noted, they used the same routes ("arboreal pathways") repeatedly during the sample period. These pathways connected trees or groups of trees that were visited repeatedly and used a number of times as food sources. It appeared that the monkeys were following the pathways to get to important potential food sources and were monitoring such trees, so that they often knew when food would be available.

As evidence that they were visiting such trees deliberately, it was noted that in the Lutz Ravine the study troop passed through individuals of *Ficus yoponensis* and *F. insipida* more frequently than expected if they had been traveling simply at random. Since *F. yoponensis* and *F. insipida* contribute both fruit and leaves to the howler diet, typically have more than one fruit crop per annum and are almost continuous flushers, there should have been a relatively high incentive for traveling to them.

Further, on some days animals in both study areas would pass through three or more trees of a given species (e.g., *Pseudobombax septenatum Platypodium elegans*) located in different parts of their supplying area and harvest the same food item from all of them. This suggests that not only do the monkeys know the locations of different food trees in their supplying area but also that they recognize that if one individual of a given species has seasonal food items, other individuals of the species are likely to have them as well.

SIZE OF SUPPLYING AREA. If food is patchy in both space and time, the feeder is likely to require a larger supplying area than if food (at the same average density) were uniformly dispersed and temporarily constant (Schoener 1971). Thus, by depending mainly on seasonal foods, howlers on Barro Colorado probably require a larger supplying area than if they depended mainly on perennial foods. Basically, the area must be large enough to include enough individuals of enough different species with enough different phenological strategies and/or enough individuals of species that are phenologically asynchronous so that there will always be a sufficient supply of seasonal foods. Further, a troop of howlers is likely to require a larger supplying area than a solitary howler (Milton and May 1976). Since having a larger supplying area is likely to require more travel, the costs of procuring food in such an area are likely to be higher. If the foraging strategy of howlers on Barro Colorado is as hypothesized, they should have evolved behavioral and/or morphological features that minimize these costs.

The size of the supplying area was almost the same in both study areas, despite the differences in forest structure and composition. It was pointed out that this area was about equal to the square of a distance that was well within the range of the distance traveled in one day by the study troops. It was suggested that the supplying area of a howler troop might not be much larger than an area that could be traversed in a day's travel, and that this might be a limiting factor on troop size, as indicated by the data from Carpenter's censuses (1934, 1962) and my observations of an unusually large howler troop. Whatever the mechanisms are, troop size does appear to be regulated and thus the required supplying area does not exceed a certain limit.

SELECTIVE USE OF SUPPLYING AREAS. Both study troops used their supplying areas very selectively, using only about 40 percent of the total area and heavily using only several hectares. The areas used more heavily than others appeared to have higher densities of preferred food species. This was demonstrated in the Lutz Ravine with respect to *Ficus yoponensis* and *F. insipida*. The selective use of areas where the probabilities of finding particular preferred foods were relatively high should have increased the productivity of travel.

OVERLAP OF SUPPLYING AREAS. The benefits and costs of defending food resources from conspecifics have been examined by Brown (1964), Brown and Orians (1970) and Schoener (1971). The benefits are in having exclusive use of food resources (with respect to conspecifics), and the costs are in patrolling the territory for invaders and repelling them. In Schoener's model (1971), at high enough levels of food density it will not be profitable to defend a territory, and at low enough levels of food density, under some circumstances, it will not be profitable either. Since the food sources of howlers on Barro Colorado are very patchy in both space and time (i.e., there is usually either more than enough of a particular item or none at all), the theory predicts that they will not defend their supplying areas—and they do not. Instead, adjacent howler troops allow their supplying areas to overlap and apparently keep one another at a distance by howling to announce their location. They almost always howl at dawn and often upon arriving in a food tree after moving into a new area and also late in the afternoon. Thus they do not incur the costs of territorial defense, while enjoying some of the benefits.

LEVEL OF ACTIVITY. Howlers spent only 26.47 percent of the day foraging (feeding and traveling). This suggests that despite the problems of food procurement, they were able to meet their requirements for energy and nutrients in little more than three hours of foraging per day. Thus, if time spent at an activity is indicative of the energy expended in the activity, they spent relatively little energy in foraging. On the other hand, they spent 65.54 percent of the day resting, which should have conserved energy and therefore reduced the *need* for energy. It may be, then, that the low level of activity of howlers is an integral part of their foraging strategy.

SOCIAL ORGANIZATION. Certain features of social organization may be viewed as part of the howler foraging strategy. As discussed above, the size of howler troops on Barro Colorado appeared to be limited, possibly by food supply factors; however, since any size group is likely to require a larger supplying area than a solitary animal, the travel cost per animal of foraging in a group is likely to be higher. The benefits of being in a group may have little or nothing to do with foraging (e.g., defense against predators, accessibility of mates); but there may also be benefits of foraging in a group, e.g., greater ability to defend a supplying area (Schoener 1971), greater efficiency in covering an area (Altmann and Altmann 1974), greater knowledge of food sources (Ripley 1970; Geist 1971; Kummer 1971), and conspecific cuing (Kiester and Slatkin 1974; Milton and May 1976). Since howler troops do not defend their supplying areas, they would not get the first benefit; and since they usually travel in single file, they would not benefit much from being able to cover a wider swath. If howlers have an advantage in foraging as a group, it probably comes from having a larger pool of information as to where and when preferred foods might be available, and how best to get them (Milton 1979b).

It may also be that where food sources are very patchy there is no disadvantage in foraging as a group, up to a certain size. This should be true if there is enough food for more than one animal in a typical patch. The patches (i.e., the canopy trees) that provided most of the foods eaten by howlers on Barro Colorado generally seemed to have enough food for more than one howler—when there was food available in the patch. So, up to a certain size, a group of howlers might not require a larger supplying area than a solitary howler, might not have to travel more and might therefore enjoy the other benefits of being in a group without any disadvantage.

Troop composition on Barro Colorado is typically 3 to 4 adult males, 5 to 7 adult females and 5 to 7 immature animals. This composition ensures that mating partners are always close at hand so that no energy need be expended in seeking them out. In addition, such troop composition has apparently led to a type of division of labor between male and female howlers. This division of labor appears to conserve energy of female howlers, which are almost continuously facing the demands imposed on them by pregnancy and lactation. The latter part of pregnancy can increase nutritional requirements by some 25 percent while lactation can increase them by 50 percent (FAOUN 1950; Portman 1972). For this reason, female howlers may often have a harder time meeting their energy requirements than males. Thus any essential behaviors that could be performed by males would be beneficial to females. I noted that males did all of the powerful howling and that only males took part in inter-troop encounters; the females remained behind and rested or fed while the males howled, chased one another, or fought. Further, in both study troops an older male appeared to coordinate most troop movement and thus determined travel routes and use of supplying area. These predominately male activities, while protecting rights to females and investments in offspring, helped the troop to locate food and to feed without interference from other troops. Females were spared involvement in these essential activities and should therefore have been able to invest more energy in reproductive activities.

MORPHOLOGY. Certain features of morphology may also be viewed as part of the howler foraging strategy.

Body Size: Howler monkeys, which weigh between 7 and 9 kg as adults, are among the largest of the Neotropical primates (Hill 1962). There are decided advantages in a relatively large body size as measured by the kilocalories required per unit weight for basal maintenance. Surface-to-volume ratio decreases as animals become larger, which in turn reduces heat loss and lowers metabolic costs (Kleiber 1961). Thus a relatively large body size would fit into a strategy of energy conservation.

Hindgut Enlargement: Though howlers lack the extensive hindgut enlargement characteristic of many primate folivores, data presented in table 1.1 show that the howler hindgut (caecum + colon) is somewhat larger than various other non-leaf-eating primates. This enlargement should permit howlers to retain quantities of ingesta in the hindgut for longer periods of time than animals lacking this additional volume. This in turn should increase fermentation efficiency. Further, the greater surface area of the howler hindgut provides more scope for the absorption of fermentation end products than otherwise would be the case. Thus the size and length of the howler hindgut can be viewed as improving net returns from foraging by increasing fermentation efficiency and the absorption of fermentation end products.

Laryngeal Specializations: The greatly enlarged hyoid bone and attendant laryngeal modifications enable howlers to produce their distinctive sonorous vocalizations. As discussed above, these vocalizations appear to function to announce the location of a troop and thereby inhibit other troops from coming into the same area to feed, to reduce the probability of inadvertent inter-troop contact, and to settle most inter-troop contacts primarily by means of vocal, rather than physical, battles. Thus the laryngeal specializations of howlers would also help to conserve energy.

Prehensile Tail: The prehensile tail of howlers appears to serve a dual function to reduce locomotor costs and to increase the efficiency of food harvesting. In traveling, by using the prehensile tail as an additional appendage, the howler is able to distribute its weight over a greater number of supports, including supports too far distant to be reached with either the arms or the legs. This enables howlers to use arboreal routes which could not otherwise support their weight and thus to reduce travel costs by opening more potential (i.e., more direct) routes to food sources. Further, by using the long prehensile tail in locomotion to maintain a grip on terminal branches when crossing between trees, howlers may reduce the need for leaping and other strenuous activities that would increase locomotor costs. In feeding, the prehensile tail enables howlers to hang below branches and harvest food items otherwise inaccessible. As pointed out by Grand (1972) and more recently by Mittermeier and Fleagle (1976), tail-hanging can increase the size of an animal's feeding sphere by as much as 150 percent over sitting or standing postures and hence greatly increase harvesting efficiency.

Dual Strategy

Foraging theory generally acknowledges two basic dietary strategies—the specialist, which exploits only one type of food and the generalist, which exploits more than one type of food (Schoener 1971; Cody 1974). Viewed within the trophic level of the primary consumer, howlers appear to use both strategies. At times of the year when they can procure food from more than one food category without excessive costs, howlers are dietary generalists, eating a wide variety of young leaves, fruits, and flowers. But at times when fruit is in short supply, the howler becomes more of a specialist, heavily exploiting flush leaves, including more flush leaves of particular species as well as some carefully selected mature leaves. Thus, howlers have a dual strategy, with foliage as a dietary base.

OTHER HOWLER SPECIES

Foraging Behavior

The foraging strategy of *Alouatta palliata* on Barro Colorado seems characteristic of members of this same species in other habitats (e.g., Costa Rica; Glander 1975) as well as other species of *Alouatta*. My short-term study of *A. seniculus* in a tropical evergreen forest in western Peru and of *A. caraya* in a riparian forest in northern Argentina showed that howlers in both areas had the same basic behavioral patterns as *A. palliata* with respect to food selection and procurement. As shown in table 7.1, regardless of species or study site, all howlers examined spent an average of 60 percent or more of their daylight hours quietly resting. In all cases a strong dependence on seasonal foods was noted and perennial resources were largely ignored. Travel to food sources generally appeared goal-directed. Except in Argentina, where howler troops lived in non-congruent forest patches, howler troops occupied overlapping home ranges and spacing between troops appeared to be effected

TABLE 7.1. Percent of Time Spent in Principal Activities in Four Howler
Study Sites

Species	Rest	Travel	Feed	Move
A. palliata				
Barro Colorado	65.54	10.23	16.24	3.16
Guanacoste, Costa Rica (from Glander 1975)	65.30	11.00	18.08	5.62
A. seniculus				
Manu, Peru	63.12	7.50	19.44	3.91
A. caraya				
Corrientes, Argentina	74.86	7.91	11.23	4.44

Note: Activity data from Barro Colorado and Guanacoste are indicative of mean
activity over an annual cycle. Data from Peru were taken over points in a 10-week
study and that from Argentina over points in a 3-week study.

by means of the howling vocalization. These results strongly suggest that all *Alouatta* species occupy the same dietary niche. This
helps to explain why howler species are found allopatrically rather
than sympatrically over their wide geographical range. In the Old
World, it is not uncommon to find members of the same primate
genus living sympatrically (e.g., *Colobus guereza* and *C. badius;
Presbytis senex* and *P. entellus*) but field studies have shown that
in such cases, each species has a distinct dietary focus.

Certain differences in foraging behavior noted between howlers
on Barro Colorado and howlers in other areas seem attributable in
large part to the different patterns of forest composition and phenology in each study area. For example, howlers in northern Argentina were less active than those in Panama, Costa Rica, or Peru and
had a much shorter day range. They were also living, at the time
of my study, on a diet composed almost entirely of the flush leaves
of only one tree species, *Inga uruguensis*. Such a diet is presumed to
be very low in ready energy. The longer periods of inactivity shown
by monkeys on this diet seem due primarily to its low energy content; the short day ranges seem due to the fact that individuals of
Inga uruguensis occurred at very high densities and little travel was
required to find food sources.

Social Organization

It is interesting to note that the different howler species studied to date appear to have somewhat different norms with respect to troop size and composition. *Alouatta palliata* in both Panama and Costa Rica live in multi-male troops composed of some 14 to 20 individuals. The adult sex ratio is around 2.5 females per male. In contrast, troops of *A. seniculus* in Peru live in single-male troops composed of some 6 individuals. The adult sex ratio here, however, is approximately the same as that of *A. palliata*. Thus far, data are insufficient to determine the average troop size of *A. caraya* but my observations to date suggest that howlers in this area tend to live in multi-male troops except in very small forest patches where single male troops may be found. Again, adult sex ratios for *A. caraya* seem similar to those of the other two *Alouatta* species I have studied. Thus group size and composition appear to vary between species and between study areas but the adult sex ratio remains relatively constant.

Students of primate behavior have long been concerned with the functional significance of different patterns of primate social organization. Since features of social organization cannot be considered apart from the total ecology of the species in question and since until recently there was a paucity of detailed ecological data available on any primate species, it is not surprising that an understanding of primate social structure has been slow to emerge. Further, in asking why species A lives in small single male troops and species B in large multi-male troops, one is ultimately asking questions related to demography. To answer such questions, life table statistics of particular primate populations are needed, as well as data on behavior and ecology. As yet, such information is not available for any *Alouatta* species and probably not for any primate species, though future research should provide it. Nevertheless, it is still possible to examine features of the social structure and ecology of different populations of *Alouatta* and discern some factors which may be involved in setting limits to group size and composition for a given population in a given habitat.

Both *Alouatta palliata* on Barro Colorado and *A. seniculus* in Peru were studied in evergreen tropical forest. Very broadly, climatic features between the two areas are similar in that there is a long rainy season, followed by a short transition season and moderate dry season. In Peru, however, the forest is characterized by a higher species diversity than that on Barro Colorado as well as more stems per unit area (Milton and Janson, unpub. data). It could therefore provide more potential food sources for howlers than Barro Colorado. Conversely, a high species diversity and many stems per unit area could provide less potential food for monkeys as evidenced by the Dipterocarp forests of Malasia. But since the Peruvian forest appears similar in its floristic composition and phenology to that on Barro Colorado, it does appear to be a potentially richer environment. Given this, it is difficult to understand why howlers in a "rich" environment would live in smaller groups than those in a "less rich" environment, all else being equal. In fact, of course, all else is not equal. Though the Peruvian forest is floristically richer, it is also faunistically richer. On Barro Colorado, howlers share the forest with four other monkey species; in Peru, howlers share the forest with eleven other monkey species. Other non-primate primary consumers are also more numerous in Peru. Thus though howlers in Peru may be said initially to have a "bigger pie," their share of this pie, after it is distributed among the many other primary consumers in the habitat, may be far lower per unit area than on Barro Colorado. There is evidence to support the view that howlers in Peru require a larger supplying area per individual than those in Panama, which would suggest that resources are less concentrated for howlers in Peru. In Peru some 6 howler monkeys require around 25 hectares of forest as a supplying area while in Panama some 18 individuals require only 31 hectares. The energetically conservative life style of howlers is predicated on living in a home range area that can be traversed in a single day and is amenable to a program of constant monitoring. In Peru, a troop size larger than some 6 individuals might be counterproductive since a concomitantly larger home range might be required (Milton and May 1976) and monitoring efficiency might be proportionately decreased. This would lower overall foraging efficiency. Thus in such a habitat, small single male troops may therefore be best adapted for dealing efficiently with patterns of resource distribution. In Panama, *A. palliata* appear to be under somewhat less rigorous pressure with respect to nonspecific competitors. Proportionately more animals can live per unit area. What this implies is that here *intraspecific* competition may be somewhat more rigorous than in Peru. Field data indicate that troops of *A. palliata* on Barro Colorado come into proximity far more often than troops of *A. seniculus* in my Peruvian study site. This might select for a multi-male troop structure to better defend the rights of reproductive partners and offspring to preferred food sources in a system of overlapping home ranges. Since sex ratios of adults are the same in Panama as in Peru, the addition of other males should not detract from the fitness of the dominant adult male and might improve the ability of the troop to secure preferred food sources. Thus a multi-male troop structure might be favored in the Barro Colorado forest over a single-male troop. Predation pressures in Panama may also be such that larger troop size is not a disadvantage.

This same type of comparative analysis can also be applied to howlers living in other areas. For example, *Alouatta palliata* troops in Guanacoste, Costa Rica, are generally smaller than those of *A. palliata* on Barro Colorado and also have a smaller home range (Glander 1975). The Guanacoste forest, which is riparian, is far less dense and diverse than the forest on Barro Colorado and is characterized by stronger seasonality. If troops were larger, home range areas would probably also be larger (Milton and May 1976). This might lower foraging efficiency, as noted above, since it would require monitoring a larger area. Further, since seasonality is much more pronounced in Costa Rica, animals here face sharper peaks and valleys in the production of seasonal dietary items and often live for relatively long periods of time on diets high in foliage compared with those on Barro Colorado (Glander 1975). Low energy diets and large home ranges seem incompatible in theory. It might be asked why troops are as large as they are; why, for example, don't howlers in this relatively depauperate habitat live in small one-male groups similar to those of *A. seniculus* in Peru? Larger troops with a multi-male structure might be advantageous here for two reasons. First, there are no other primate primary consumers living in the Guanacoste study site, which implies that howlers get a larger share of the available resources than otherwise. The manner in which the resources are distributed in space and time might maintain some 12 to 15 howlers as easily as 6 and thus a larger troop size, up to a point, is not a disadvantage. Further, a multi-male troop structure may aid in defending the rights of particular troops to resources. Again since the adult sex ratio remains the same as in single-male troops, the dominant male suffers no disadvantage here from the presence of other males.

At this stage of research, much of the above must be regarded as speculation. However, in time the collection of a wide sample of patterns of howler social organization and demography as well as data on specific features of the habitats (such as density, distribution, and phenological patterns of the trees, other primary consumers in the habitat, predators, and the like) should enable researchers to test these hypotheses.

OTHER LEAF-EATING PRIMATES

Howlers are not unique in being able to exploit foliage as a principal food source without extensive digestive specializations such as those of colobines and indriids. A number of other primates that appear to lack such pronounced specializations also eat considerable amounts of foliage. This list includes some lemurs (e.g., *Lemur fulvus rufus*, *Lemur catta*; Sussman 1972; *Hapalemur griseus*; Petter and Peyrieras 1970), some cercopithecine monkeys (e.g., *Cercopithecus ascanius*, Haddow 1965; *C. mitis*; Rudran 1978), lowland gorillas (*Gorilla gorilla gorilla*; Sabater Pi 1974), mountain gorillas (*Gorilla gorilla berengei*; Schaller 1965; Casimir 1975; Fossey and Harcourt 1977; Goodall 1977), siamangs (*Symphalangus syndactylus*; Chivers 1974) and at least two other genera of New World monkey, *Brachyteles arachnoides* (Milton, unpub.) and *Lagothrix lagotricha* (Kay 1974; Kavanagh and Dresdale 1975; Nishimura, pers. comm.). As leaves generally are high in structural carbohydrates and low in ready energy components, all of these primates—to a greater or lesser extent, depending on food density and distribution—should face the same basic problems of food digestion, food selection, and food procurement. Thus, they should have behavioral and morphological adaptations for selecting foods of a relatively high quality while minimizing the costs of procuring such foods and/or conserving energy.

The available data indicate these primates are rather selective in their feeding and that most of them eat fruit and flowers as well as foliage. *Lemur catta* and *L. fulvus rufus* both have substantial proportions of fruit in their diets in certain areas and at certain times of the year (Sussman 1972). *Hapalemur griseus* specialize on the shoots of bamboos and reeds (Petter and Peyrieras 1970) and appear to have specialized dentition for harvesting such a diet (Milton 1978). Both Schaller (1965) and Casimir (1975) emphasized that mountain gorillas are highly selective in the parts and species of plants they eat, and Sabater Pi (1974) found that the lowland gorilla has a substantial proportion (40 percent) of fruit in the diet.

The Malaysian siamangs studied by Chivers (1974) have a diet that appears to be very similar to that of the howlers on Barro Colorado. They eat similar proportions of young leaves and fruits, particularly fig fruit. Indeed, like the howlers in the Lutz Ravine, they spend about half of their feeding time on fig products. There are, however, some important behavioral differences between howlers and siamangs. Siamangs live in pair-bonded family groups, are territorial, and spend about twice as much time foraging. But siamangs are about twice the size of howlers and, where studied by Chivers (1974), live in a habitat with less seasonality than on Barro Colorado. As pointed out by Curtin and Chivers (1978), the diet of siamangs shows an unusual degree of consistency throughout the year; the proportion of time spent on each dietary category remains more or less constant at all times in an annual cycle—in striking contrast to the diet of howlers in Panama. Since foods are less patchy in time, siamangs should be more likely to defend a territory; and since they are large primates, they can repel any other primate competitors from potential food sources (Chivers 1974). Also, siamangs have a locomotor mode, brachiation, that should be less costly per unit distance than that of howlers, which would reduce travel costs in both foraging and territorial defense.

Further comparisons are hampered by the lack of data on forest structure and phenology, food content, assimilation efficiencies, and energy expenditure, among other variables. It is hoped that in the future these aspects will be examined in greater detail and that eventually such information can be extended to include other members of particular primate genera living in different habitats. In time, a comprehensive model of primate foraging strategies may thus be constructed and integrated into a general theory of foraging strategy.

REFERENCES

Allison, M. J. 1965. "Nutrition of Rumen Bacteria." In R. W. Dougherty et al., eds. *Physiology of Digestion in the Ruminant*, pp. 369-78. Washington, D.C.: Butterworths.

Altmann, J. 1974. "Observational study of behavior; sampling methods." *Behavior*, 69(3-4):227-67.

Altmann, S. A. 1959. "Field observations on a howling monkey society." *J. Mammal.* 40:317-30.

——. 1970. "Baboons, space, time, and energy." *Amer. Zool.*, 14:221-48.

Altmann, S. A. and J. Altmann. 1970. *Baboon Ecology*. Chicago: University of Chicago Press.

Augspurger, C. K. 1978. "Reproductive Consequences of Flowering Synchrony in *Hybanthus prunifolius* (Violaceae) and Other Shrub Species of Panama." Ph.D. dissertation, University of Michigan, Ann Arbor.

Baldwin, J. D. and J. I. Baldwin. 1972. "Population density and use of space in howling monkeys (*Alouatta villosa*) in southwestern Panama." *Primates*, 13:371-79.

Bauchop, T. 1978. "Digestion of Leaves in Vertebrate Arboreal Folivores." In G. G. Montgomery, ed., *The Ecology of Arboreal Folivores*, pp. 193-204. Washington, D.C.: Smithsonian Press.

——. 1971. "Stomach microbiology of primates. *Ann. Review of Microbio.*, 25:429-36.

Bauchop, T. and R. W. Martucci. 1968. "Ruminant-like digestion of the langur monkey." *Science*, 161:698-700.

Bell, E. A. 1972. "Toxic Amino Acids in the Leguminosae." In J. B. Harborne, ed., *Phytochemical Ecology*, pp. 163-78. New York: Academic Press.

Bell, R. H. V. 1971. "A grazing ecosystem in the Serengeti." *Sci. Amer.*, 225:86-93.

Bernstein, I. S. 1964. "A field study of the activities of howler monkeys." *Anim. Behav.*, 12:92-97.

Blackburn, T. H. 1965. "Nitrogen Metabolism in the Rumen. In R. W. Dougherty et al., eds., *Physiology of Digestion in the Ruminant*, pp. 322-31. Washington, D.C.: Butterworths.

152   References

Brown, J. L. 1964. "The evolution of diversity in avian territorial systems." *Wilson Bull.*, 6:160–69.

Brown, J. L. and G. H. Orians. 1970. "Spacing patterns in mobile animals." *Ann. Review of Ecol. and System.*, 1:239–62.

Burt, W. H. 1943. "Territoriality and home-range concepts as applied to mammals." *J. Mammal.*, 24:346–52.

Carpenter, C. R. 1934. "A field study of the behavior and social relations of howling monkeys." *Comp. Psychol. Monog.*, 10, 48:1–168.

———. 1962. "Field Studies of a Primate Population. In E. L. Bliss, ed., *Roots of Behavior*, pp. 286–94. New York: Harper.

Casimir, N. J. 1975. "Feeding ecology and nutrition of an eastern gorilla group in the Mt. Kahuzi Region (Republic de Zaire)." *Folia Primatol.*, 24:81–136.

Chivers, D. 1969. "On the daily behavior and spacing of howler monkey groups." *Folia Primatol.*, 10:48–102.

———. 1974. *The Siamang in Malaya: A Field Study of a Primate in Tropical Rain Forest*. Basel: Karger.

Clutton-Brock, T. H. 1974. "Primate social organization and ecology." *Nature*, 250:539–42.

———. 1975. "Feeding behavior of red colobus and black and white colobus in East Africa." *Folia Primatol.*, 23:165–207.

Cody, M. L. 1974. "Optimization in ecology." *Science*, 183:1156–64.

Coelho, A. M., C. A. Bramblett, L. B. Quick, and S. S. Bramblett. 1976. "Resource availability and population density in primates: A socio-bioenergetic analysis of the energy budget of Guatemalan howler and spider monkeys." *Primates*, 17:63–80.

Collias, D. E. and C. H. Southwick. 1952. "A field study of the population density and social organization in howler monkeys." *Proc. Amer. Phil. Soc.*, 96:144–56.

Conover, W. J. 1971. *Practical Nonparametric Statistics*. New York: Wiley.

Cramer, D. L. 1968. "Anatomy of the Thoracic and Abdominal Viscera." In M. R. Malinow, *Biology of the Howler Monkey*, pp. 90–103. Basel: Karger.

Croat, T. B. 1967. "Seasonal flowering in central Panama." *Ann. Missouri Botan. Gard.*, 56:295–307.

———. 1978. *Flora of Barro Colorado Island*. Palo Alto, Calif.: Stanford University Press.

Curtin, S. H. and D. J. Chivers. 1978. "Leaf-Eating Primates of Peninsular Malaysia: The Siamang and Dusky Leaf Monkey." In G. G. Montgomery, ed., *The Ecology of Arboreal Folivores*, pp. 441–64. Washington, D.C.: Smithsonian Press.

Ehrlich, P. R. and P. H. Raven. 1965. "Butterflies and plants: A study in co-evolution." *Evol.*, 18:586–608.

Eisenberg, J. F., N. A. Muckenhirn, and R. Rudran. 1972. "The relation between ecology and social structure in primates." *Science*, 176:863–74.

Eisenberg, J. F. and R. W. Thorington. 1973. "A preliminary analysis of Neotropical mammal fauna." *Biotropica*, 5:150-61.

Ellefson, J. O. 1968. "Territorial Behavior in the Common White-handed Gibbon, *Hylobates lar* Linn." In P. C. Jay, ed., *Primates: Studies in Adaptation and Variability*, pp. 180-99. New York: Holt, Rinehart & Winston.

Emlen, J. M. 1966. "The role of time and energy in food preference." *Amer. Nat.*, 100:611-17.

FAUON. 1950. "Calorie requirements." Washington, D.C.: F.A.O. Publications.

Feeny, P. 1970. "Seasonal changes in oak leaf tannins and nutrients as a cause of spring feeding by winter moth caterpillars." *Ecology*, 51:565-80.

Flower, W. H. 1892. "Comparative anatomy of the organs of digestion of the Mammalia." *Medical Times & Gazette*, 1:561-67.

Fogden, M. P. L. 1972. "The seasonality and population dynamics of equatorial forest birds in Sarawak." *Ibis*, 114:307-43.

Fooden, J. 1964. "Stomach contents and gastro-intestinal proportions in wild-shot Guianan monkeys." *Amer. J. Phys. Anthrop.*, 22:227-32.

Fossey, D. and A. H. Harcourt. 1977. "Feeding Ecology of Free-Ranging Mountain Gorillas (*Gorilla gorilla beringei*)." In T. H. Clutton-Brock, ed., *Primate Ecology*, pp. 415-49. London: Academic Press.

Foster, R. B. 1973. "Seasonality of Fruit Production and Seed Fall in a Tropical Forest Ecosystem in Panama." Ph.D. thesis, Duke University, Durham, N.C.

Fowden, L. 1974. "Nonprotein Amino Acids from Plants; Distribution, Biosynthesis, and Analog Functions." In V. C. Runeckles and E. E. Conn, eds., *Metabolism and Regulation of Secondary Plant Products*, pp. 95-122. New York: Academic Press.

Fraenkel, G. 1959. "The raison d'etre of secondary plant substances." *Science*, 129:1466-70.

Frankie, G. W., H. G. Baker, and P. A. Opler. 1974. "Comparative phenological studies of trees in tropical wet and dry forests in the lowlands of Costa Rica." *J. Ecology*, 62:881-919.

Freeland, W. J. and D. H. Janzen. 1974. "Strategies of herbivory in mammals; the role of plant secondary compounds." *Amer. Nat.*, 108:269-89.

Geist, V. 1971. *Mountain Sheep: A Study in Behavior and Evolution*. Chicago: University of Chicago Press.

Glander, K. E. 1975. "Habitat and Resource Utilization: An Ecological View of Social Organization in Mantled Howler Monkeys." Ph.D. dissertation, University of Chicago.

Goodall, A. G. 1977. "Feeding and Ranging Behaviour of a Mountain Gorilla Group (*Gorilla gorilla beringei*) in the Tshibinda-Kahuzi Region (Zaire)." In T. H. Clutton-Brock, ed., *Primate Ecology*, pp. 450-80. London: Academic Press.

Grand, T. I. 1972. "A mechanical interpretation of terminal branch feeding." *J. Mammal.*, 53:198-201.

Greig-Smith, P. 1964. *Quantitative Plant Ecology*. New York: Plenum Press.

Guthrie, H. A. 1971. *Introductory Nutrition*. St. Louis: C. V. Mosby.

Haddow, A. J. 1952. "Field and laboratory studies on an African monkey, *Cercopithecus ascanius schmidti* Matschie." *Proc. Zool. Soc. Lond.*, 122:297–394.

Hankioja, E. and P. Niemala. 1976. "Does birch defend itself against herbivores?" *Rep. Kevo Subarctic Res. Stat.*, 13:44–47.

Hegnauer, R. 1963. *Chemotaxonomie der Pflanzen*, vols. 1–4. Basel: Berkhauser.

Hershkovitz, P. 1969. "The evolution of mammals on Southern continents, VI: The recent mammals of the Neotropical region, a zoogeographic and ecological review." *Quart. Rev. Biol.*, 44:1–70.

Hill, W. C. O. 1953. *Primates: Comparative Anatomy and Taxonomy*, vol. 1. New York: Intersciences.

———. 1962. *Primates: Comparative Anatomy and Taxonomy*, vol. 4. New York: Intersciences.

Hladik, A. and C. M. Hladik. 1969. "Rapports trophiques entre végétation et primates dans la forêt de Barro Colorado (Panama)." *Terre et Vie*, 1:25–117.

Hladik, C. M. 1967. "Surface relative du tractus digestif de quelques primates. Morphologie des villosités intestinales et correlations avec le régime alimentaire." *Mammalia*, 31:120–47.

———. 1977a. "A Comparative Study of the Feeding Strategies of Two Sympatric Species of Leaf Monkey: *Presbytis senex* and *Presbytis entellus*. In T. H. Clutton-Brock, ed., *Primate Ecology*, pp. 324–54. London: Academic Press.

———. 1977b. "Chimpanzees of Gabon and Chimpanzees of Gombe: Some Comparative Data on the Diet." In T. H. Clutton-Brock, ed., *Primate Ecology*, pp. 481–503. Washington, D.C.: Smithsonian Press.

———. 1978. "Adaptive Strategies of Primates in Relation to Leaf-Eating." In G. G. Montgomery, ed., *The Ecology of Arboreal Folivores*, pp. 373–95. Washington, D.C.: Smithsonian Press.

Hladik, C. M. and A. Hladik, 1972. "Disponibilités alimentaires et domaines vitaux des primates à Ceylon." *Terre et Vie*, 2:149–215.

Hladik, C. M., A. Hladik, J. Bousset, P. Valdebouze, G. Viroben, and J. Delort-Laval. 1971. "Le regime alimentaire des primates de l'ile de Barro Colorado (Panama)." *Folia Primatol.*, 16:85–122.

Hungate, R. E. 1967. "Ruminal Fermentation." In C. F. Cole, ed., *Handbook of Physiology*, 5:2725–45. Baltimore: Waverly Press.

Huntsburger, D. V. and P. Billingsley. 1973. *Elements of Statistical Inference*. Boston: Allyn & Bacon.

Jackson, J. F. 1978. "Seasonality of flowering and leaf-fall in a Brazilian subtropical lower montane moist forest." *Biotropica*, 10:38–42.

Janzen, D. H. 1967. "Synchronization of sexual reproduction of trees within the dry season in Central America." *Evolution*, 21:620–37.

_____. 1969. "Seed-eaters versus seed size, number, dispersal, and toxicity." *Evolution*, 23:1-27.

_____. 1971. "Seed predation by animals." *Ann. Rev. Ecol. Syst.*, 2:465-92.

Jay, P. 1965. "The Common Langur of North India." In I. DeVore, ed., *Primate Behavior*, pp. 197-249. New York: Holt, Rinehart & Winston.

Jewell, P. A. 1966. "The concept of home range in mammals." *Symp. Zool. Soc. Lond.*, 18:85-110.

Jolly, A. 1972. *The Evolution of Primate Behavior*. Chicago: Chicago University Press.

Kavanagh, M. and L. Dresdale. 1975. "Observations on the woolly monkey (*Lagothrix lagothricha*) in northern Colombia." *Primates*, 16:285-94.

Kay, R. F. 1974. "Mastication, Molar Tooth Structure, and Diet in Primates." Ph.D. dissertation. Yale University, New Haven, Conn.

Kershaw, K. A. 1973. *Quantitative and Dynamic Plant Ecology*. London: William Clowes.

Keys, J. E., Jr., P. J. Van Soest, and E. P. Young. 1969. "Comparative study of the digestibility of forage cellulose and hemicellulose in ruminants and nonruminants." *J. Animal Science*, 29:11-15.

Kiester, A. R. and M. Slatkin. 1974. "A strategy of movement and resource utilization." *Theoretical Population Biology*, 6:1-20.

Klieber, M. 1961. *The Fire of Life*. New York: Wiley.

Knight, D. H. 1975. "A phytosociological analysis of species-rich tropical forest on Barro Colorado Island, Panama." *Ecological Monographs*, 45:259-84.

Kummer, H. 1971. *Primate Societies*. Chicago: Aldine Press.

Lang, G. E. 1969. "Sampling Tree Density with Quadrats in a Species-Rich Tropical Forest." M.A. thesis, University of Wyoming, Laramie.

Leigh, E. G. 1975. "Structure and climate in tropical rain forest." *Ann. Review of Ecol. and System.*, 6:67-85.

Leigh, E. G. and N. Smythe. 1978. "Leaf Production, Leaf Consumption, and the Regulation of Folivory on Barro Colorado Island." In G. G. Montgomery, ed., *The Ecology of Arboreal Folivores*, pp. 33-50. Washington, D.C.: Smithsonian Press.

Levin, D. A. 1971. "Plant phenolics: an ecological perspective." *Amer. Nat.*, 105:157-81.

_____. 1976. "Alkaloid-bearing plants: an ecological perspective." *Amer. Nat.*, 110:261-84.

MacArthur, R. and E. R. Pianka. 1966. "On optimal use of a patchy environment." *Amer. Nat.*, 100:603-9.

McKey, D. 1974. "Adaptive patterns in alkaloid physiology." *Amer. Nat.*, 108:305-20.

_____. 1975. "The Ecology of Coevolved Seed Dispersal Systems." In L. E. Gilbert and P. H. Raven, eds., *Coevolution of Plants and Animals*, pp. 159-91. Austin: University of Texas Press.

McKey, D., P. G. Waterman, C. N. Mbi, J. S. Gartlan, and T. T. Struhsaker.

1978. "Phenolic content of vegetation in two African rain forests: Ecological implications." *Science*, 202:61–64.

Marler, P. 1970. "Vocalizations of East African monkeys: I, Red colobus." *Folia Primatol.*, 13:81–91.

Mendel, F. 1976. "Postural and locomotor behavior of *Alouatta palliata* on various substrates." *Folia Primatol.*, 26:36–53.

Milton, K. 1975. "Urine-rubbing behavior in the mantled howler monkey." *Folia Primatol.*, 23:105–12.

———. 1977. "Howler monkey population on Barro Colorado at 1,300." *Star & Herald*, Panama, R.P., April 17, 1977, p. 1.

———. 1978a. "Role of the upper canine and p^2 in increasing the harvesting efficiency of *Hapalemur griseus* Link 1795." *J. Mammal.*, 59:188–90.

———. 1978b. "Behavioral Adaptations to Leaf-Eating by the Mantled Howler Monkey." In G. G. Montgomery, ed., *The Ecology of Arboreal Folivores*, pp. 535–50. Washington, D.C.: Smithsonian Press.

———. 1978c. "The Quality of Diet as a Possible Limiting Factor on the Barro Colorado Island Howler Monkey Population." In *Proceedings of the Sixth Congress of the International Primate Society*. Cambridge: Academic Press.

———. 1979a. "Factors influencing leaf choice by howler monkeys: a test of some hypotheses of food selection by generalist herbivores." *Amer. Nat.*, 114:362–78.

———. 1979b. "The spatial and temporal patterns of plant foods in tropical forests as a stimulus to intellectual development in primates." Abstract. *Amer. J. Phys. Anthrop.*, 50:464–65.

Milton, K. and M. May. 1976. "Body weight, diet, and home-range area in primates." *Nature*, 259:459–62.

Milton, K., T. M. Casey, and K. K. Casey. 1979. "The basal metabolism of mantled howler monkeys." *J. Mammal.*, 60:373–76.

Mittermeier, R. A. 1973. "Group activity and population dynamics of the howler monkey on Barro Colorado Island." *Primates*, 14:1–19.

Mittermeier, R. A. and J. G. Fleagle. 1976. "The locomotor and postural repertoires of *Ateles geoffroyi* and *Colobus guereza*, and a reevaluation of the locomotor category semibrachiation." *Amer. J. Phys. Anthrop.*, 45:235–56.

Moir, R. J. 1965. "The Comparative Physiology of Ruminantlike Animals." In R. W. Dougherty et al., eds., *Physiology of Digestion in the Ruminant*, pp. 1–14. Washington: Butterworths.

———. 1967. "Ruminant Digestion and Evolution." In C. F. Cole, ed., *Handbook of Physiology*, 5:2673–94. Baltimore: Waverly Press.

Morrison, D. W. 1975. "The Foraging Behavior and Feeding Ecology of a Neotropical Fruit Bat, *Artibeus jamaicensis*." Ph.D. dissertation, Cornell University, Ithaca, N.Y.

———. 1978. "Foraging ecology and energetics of the frugivorous bat *Artibeus jamaicensis*." *Ecology*, 59:716–23.

Morton, E. S. 1973. "On the evolutionary advantages and disadvantages of fruit eating in tropical birds." *Amer. Nat.*, 107:8-22.

Murton, R. K., A. J. Isaacson, and N. J. Westwood. 1966. "The relationship between wood pigeons and their clover food supply and the mechanisms of population control." *J. Applied Ecology*, 3:55-96.

Nagy, K. A. and K. Milton. 1979a. "Aspects of dietary quality, nutrient assimilation, and water balance in wild howler monkeys." *Oecologia*, 39: 249-58.

———. 1979b. "Energy metabolism and food consumption by wild howler monkeys." *Ecology*, 60:475-80.

Neville, M. K. 1972. "The population structure of red howler monkeys (*Alouatta seniculus*) in Trinidad and Venezuela." *Folia Primatol.* 17: 56-86.

Parke, D. V. 1968. *The Biochemistry of Foreign Compounds*. Oxford: Pergamon Press.

Parra, R. 1978. "Comparison of Foregut and Hindgut Fermentation in Herbivores." In G. G. Montgomery, ed., *The Ecology of Arboreal Folivores*. Washington, D.C.: Smithsonian Press.

Petter, J. J. and A. Peyrieras. 1970. "Observations ecoethologiques sur les Lemuriens malgaches du genre *Hapalemur*." *Terre et Vie*, 24:356-82.

Pielou, E. C. 1969. *An Introduction to Mathematical Ecology*. New York: Wiley.

Portman, O. W. 1972. "Nutrient Requirements of the Monkey." In *Nutrient Requirements of Laboratory Animals*, no. 10. Washington, D.C.: N.S.A. Publications.

Puerto, A., J. A. Deutsch, F. Molina, and P. L. Roll. 1976. "Rapid discrimination of rewarding nutrients by the upper gastrointestinal tract." *Science*, 192:485-87.

Rhoades, D. F. and R. G. Cates. 1976. "Toward a General Theory of Plant Antiherbivore Chemistry." In J. W. Wallace and R. L. Mansell, eds., *Biochemical Interaction Between Plants and Insects*, pp. 168-213. New York: Plenum.

Reichman, O. J. 1975. "Relation of desert rodent diets to available resources." *J. Mammal.*, 56:731-51.

Richard, A. 1970. "A comparative study of the activity patterns and behavior of *Alouatta villosa* and *Ateles geoffroyi*." *Folia Primatol.*, 12: 241-63.

Richard, A. F. 1977. "The Feeding Behaviour of *Propithecus verreauxi*." In T. H. Clutton-Brock, *Primate Ecology*, pp. 72-96. London: Academic Press.

Richards, P. W. 1952. *The Tropical Rain Forest*. Cambridge: Cambridge University Press.

Ripley, S. 1970. "Leaves and Leaf-Monkeys." In J. R. Napier and P. H. Napier, eds., *Old World Monkeys*, pp. 481-509. New York: Academic Press.

Rockwood, L. L. 1974. "Seasonal changes in the susceptibility of *Crescentia slata* leaves to the flea beetle, *Oedinychus* sp." *Ecology*, 55:142–48.

Rudran, R. 1978. "Intergroup Dietary Comparisons and Folivorous Tendencies of Two Groups of Blue Monkeys (*Cercopithecus mitis stuhlmanni*). In G. G. Montgomery, ed., *The Ecology of Arboreal Folivores*, pp. 483–504. Washington, D.C.: Smithsonian Press.

Ryan, C. A. and T. R. Green. 1974. "Proteinase Inhibitors in Natural Plant Protection." In V. C. Runckles and E. E. Conn, eds., *Metabolism and Regulation of Secondary Plant Products*, pp. 123–40. New York: Academic Press.

Sabater Pi, J. 1974. "Consideraciones y comentarios sobre la alimentacion de los gorilas del africa occidental en la naturaleza." *Zoo Revista*, 21: 13–15.

Schaller, G. B. 1965. "The Behavior of the Mountain Gorilla." In I. DeVore, ed., *Primate Behavior*, pp. 324–67. New York: Holt, Rinehart & Winston.

Scheline, R. R. 1968. "Drug metabolism by intestinal microorganisms." *J. Pharmaceutical Sciences*, 57:2021–37.

Schlichte, H. G. 1978. "A Preliminary Report on the Habitat Utilization of a Group of Howler Monkeys (*Alouatta villosa pigra*) in the National Park of Tikal, Guatemala." In G. G. Montgomery, ed., *The Ecology of Arboreal Folivores*, pp. 551–60. Washington, D.C. Smithsonian Press.

Schoener, T. W. 1969. "Optimal size and specialization in constant and fluctuating environments: An energy-time approach." *Brookhaven Symp. Biol.*, 22:103–14.

——. 1971. "Theory of feeding strategies." *Ann. Review of Ecol. and System.*, 2:369–403.

Schuster, L. 1964. "Metabolism of drugs and toxic substances." *Ann. Review of Biochem.*, 33:571–96.

Sinclair, A. R. E. 1974. "The natural regulation of buffalo populations in East Africa." *E. Afr. Wildl. J.*, 12:291–311.

——. 1977. *The African Buffalo*. Chicago: University of Chicago Press.

Smith, C. C. 1977. "Feeding Behaviour and Social Organization in Howler Monkeys." In T. H. Clutton-Brock, *Primate Ecology*, pp. 97–126. London: Academic Press.

Smith, J. D. 1970. "The systematic status of the black howler monkey, *Alouatta pigra Lawrence. J. Mammal.*, 51:358–69.

Smythe, N. 1970. "Relationships between fruiting seasons and seed dispersal methods in a Neotropical forest." *Amer. Nat.*, 104:25–35.

Spencer, P. W. 1974. "Season of the apple." *Natural History*, 83:38–45.

Spiller, G. A. and R. J. Amen. 1975. "Dietary Fiber and Human Nutrition." In *Critical Reviews in Food Science and Nutrition*, pp. 39–70. Cleveland: Chemical Rubber Co.

Stern, J. T. 1971. "Functional myology of the hip and thigh of cebid monkeys and its implications for the evolution of erect posture." *Biblioth. Primatol.*, 14:1–318.

Sussman, R. W. 1972. "An Ecological Study of Two Madagascan Primates: *Lemur fulvus rufus (Audebert)* and *Lemur catta (Linnaeus).*" Ph.D. dissertation, Duke University, Durham, N.C.

Van Soest, P. 1980. *Nutritional Ecology of the Ruminant.* San Francisco: Freeman Press.

Varley, G. C. and G. R. Gradwell. 1962. "The effects of partial defoliation by caterpillars on the timber production of oak trees in England." *Proc. 11th Intern. Cong. Entomol.,* 2:211-14.

Waser, P. W. and O. Floody. 1974. "Ranging patterns of the mangabey, *Cercocebus albigena,* in the Kibale Forest, Uganda." *Z. Tierpsychol.,* 35:85-101.

Westoby, M. 1974. "An analysis of diet selection by large generalist herbivores." *Amer. Nat.,* 108:290-304.

Whittaker, R. H. and P. P. Feeny. 1971. "Allelochemics: chemical interactions between species." *Science,* 171:757-70.

Williams, R. T. 1969. *Detoxication Mechanisms.* New York: Wiley.

_____. 1971. "Species Variation in Drug Biotransformations." In B. N. LaDu et al., eds., *Fundamentals of Drug Metabolism and Drug Disposition,* pp. 187-205. Baltimore: Williams & Wilkins.

Yoshiba, K. 1968. "Local and Intertroop Variation in Ecology and Social Behavior of Common Indian Langurs." In P. Jay, ed., *Primates,* pp. 217-42. New York: Holt, Rinehart & Winston.

INDEX

Printed in the USA
CPSIA information can be obtained
at www.ICGtesting.com
JSHW021321221024
72173JS00011B/1628

9 780231 048507